Coordination Chemistry of Macrocyclic Compounds

Edwin C. Constable

Professor of Inorganic Chemistry, University of Basel,
Switzerland

Series sponsor: **ZENECA**

ZENECA is a major international company active in four main areas of business:
Pharmaceuticals, Agrochemicals and Seeds, Specialty Chemicals, and Biological Products.

ZENECA's skill and innovative ideas in organic chemistry and bioscience create products
and services which improve the world's health, nutrition, environment, and quality of life.

ZENECA is committed to the support of education in chemistry and chemical engineering.

OXFORD
UNIVERSITY PRESS

OXFORD

UNIVERSITY PRESS

Great Clarendon Street, Oxford OX2 6DP
Oxford University Press is a department of the University of Oxford
and furthers the University's aim of excellence in research, scholarship,
and education by publishing worldwide in

Oxford New York
Athens Auckland Bangkok Bogotá Buenos Aires Calcutta
Cape Town Chennai Dar es Salaam Delhi Florence Hong Kong Istanbul
Karachi Kuala Lumpur Madrid Melbourne Mexico City Mumbai
Nairobi Paris São Paulo Singapore Taipei Tokyo Toronto Warsaw

and associated companies in Berlin Ibadan

Oxford is a registered trade mark of Oxford University Press

Published in the United States
by Oxford University Press Inc., New York

British Library Cataloguing in Publication Data
Data available

Library of Congress Cataloging in Publication Data
ISBN 0–19–855692–6 (Pbk)

Typeset by the author

Printed in Great Britain
on acid-free paper by
The Bath Press, Avon

Series Editor's Foreword

Macrocycle chemistry is perhaps the most substantial innovation within coordination chemistry in recent years. From these ligands, metal-organic compounds have attained an art-form, with the trefoil knot perhaps the most elegant of these shapes. The new slant on coordination chemistry that these ligands have provided have allowed us to examine the basis of metal-ligand complexes in a new light.

Oxford Chemistry Primers are designed to give a concise introduction to all chemistry students by providing the material that would usually form an 8–10 lecture course. As well as providing up-to-date information, this series expresses the explanations and rationales that form the framework of current understanding of inorganic chemistry. Ed Constable here provides an authorative description of the basis of macrocycle coordination chemistry, ranging from ligand synthesis, metal selectivity, coordination geometries to reactivity. This book is a mine of detailed information that should be invaluable to third- or fourth-year undergraduate or for postgraduate courses in modern coordination chemistry.

John Evans
Department of Chemistry,
University of Southampton

Preface

This book deals with the coordination chemistry of macrocyclic ligands. Take the current issue of one of the major chemistry journals off the library shelf and you will be certain to find a number of articles describing the synthesis, characterisation, properties or application of a macrocyclic ligand. This book tries to explain why these ligands are used so widely and to introduce the reader to some of the unusual properties of their complexes. As in any such text, the omissions far outweigh the content and I apologise in advance to any reader whose favourite ligand system has been omitted.

The structure of the book is one that a synthetic chemist will find familiar. Ligand synthesis is followed by the synthesis of metal complexes. The properties of the metal complexes are then dealt with and the final chapter gives an introduction to the properties of macrocyclic compounds beyond the bounds of metal ion coordination chemistry. After much heart-searching, all coordinate bonds have been drawn as normal bonds without an arrowhead in order to avoid overly cluttered graphics.

I should like to thank Professor Thomas Kaden, Professor Catherine Housecroft and Dr Paul Bowyer for critical reading of the text and many useful suggestions. I also give my grateful thanks to Catherine for clarifying some of the more obscure features of OUP style sheets and typing some of the chapters. Finally, I must thank all at OUP, John Evans and Steve Davies for their patience!

Basel   E. C. C.
August 1998

Contents

1 Introduction to macrocyclic chemistry

1.1 What is a macrocycle?

This book is about macrocyclic chemistry. But what is macrocyclic chemistry? This is a very reasonable question – but like most reasonable questions it is not one which is so easy to answer. Obviously a macrocycle is a large cyclic molecule, and with the emphasis of this book and the prejudiced eyes of a coordination chemist, a macrocyclic ligand should contain donor atoms which may form coordinate bonds with metal centres. For the coordination chemist, a good working definition is that a macrocycle is a cyclic molecule with three or more potential donor atoms in a ring of at least nine atoms. By this criterion, ethylene oxide **1.1**, 1,4-dithiane **1.2**, cyclotetradecane **1.3** and cyclooctatetraene **1.4** (Fig. 1.1) are not macrocycles, whereas molecules such as cyclam **1.5**, porphyrin **1.6**, 1,4,7-trithiacyclononane **1.7** and dibenzo-18-crown-6 **1.8** fit the definition (Fig. 1.2). Although the definition is pragmatic and empirical, it essentially means that a macrocyclic ligand can bind a metal centre *within* the central cavity. If the cyclic molecule is too small, or if there are no donor atoms, it does not fit this criterion. Obviously, it is not possible to be religious in the definition. Although **1.4** contains no conventional donor atoms, it could certainly present four π-acceptor olefin sites to an electron rich metal centre.

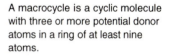

A macrocycle is a cyclic molecule with three or more potential donor atoms in a ring of at least nine atoms.

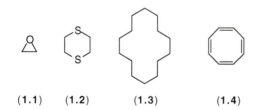

(1.1)   (1.2)   (1.3)   (1.4)

Fig. 1.1. Some cyclic molecules which are not usually thought of as macrocycles.

In this book the synthesis and coordination chemistry of macrocyclic ligands will be described. The coordination behaviour of macrocycles is no different in principle from that of open-chain polydentate ligands, although in practice, the unusual and often unexpected properties of the systems justify their being given special treatment. Some of these aspects will be developed in Section 1.3

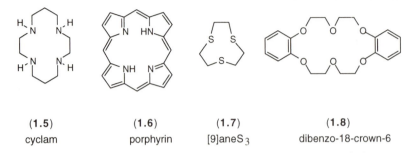

(1.5) cyclam

(1.6) porphyrin

(1.7) [9]aneS$_3$

(1.8) dibenzo-18-crown-6

Fig. 1.2. Four ligands which are usually considered to be macrocyclic.

1.2 Some nomenclature

I.U.P.A.C. = International Union of Pure and Applied Chemistry. The I.U.P.A.C. formulates rules for the nomenclature of organic and inorganic compounds. These are precise but may be unwieldy.

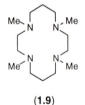

(1.9)

Nomenclature is always problematic. It is *always* possible to devise a fully systematic I.U.P.A.C. name for an organic ligand. Very often that I.U.P.A.C. name is so long and unfriendly, that, although strictly correct, it is cumbersome and no longer conveys the information in a useful manner. This is frequently the case with macrocyclic ligands. As an example, the I.U.P.A.C. names for **1.5** and **1.8** are 1,4,8,11-tetraazatetracyclodecane and 2,5,8,15,18,21-hexaoxatricyclo[20.4.0.0^{9,14}]hexacosa-1(22),8,11,13,23, 24-hexaene respectively. Not surprisingly, these complicated names are not often used and the majority of macrocyclic ligands have trivial names. Thus, **1.5** is almost universally known as cyclam (*cyclic am*ine) whilst **1.8** is called dibenzo-18-crown-6. The 'crown' nomenclature of the latter compound relates to the conformation adopted by the macrocyclic ether, which resembles a crown. Fig. 1.3 shows the conformation of the parent 18-crown-6 molecule emphasizing this structural feature. The name further conveys that the macrocyclic ring contains 18 atoms and six oxygen donors. These compounds are generically known as *crown ethers*.

Fig. 1.3. The structure of 18-crown-6 showing the 'crown-like' conformation.

(1.10)

A variety of semi-systematic methods of nomenclature have developed which are both useful and self-evident in their application. For example, compound **1.9** is *N,N',N'',N'''*-tetramethylcyclam whilst **1.10** is 15-crown-5. Another notation that we shall use is related to that used for crown ethers and indicates the size of the ring, the number and the type of donor atoms, and any substituents that are present. In this system, **1.5** is called [14]aneN$_4$ and **1.9** is 1,4,8,11-Me$_4$-[14]aneN$_4$. In general, this book will use whatever nomenclature is unambiguous and most appropriate for the compounds being discussed, rather than adopting a uniformly consistent approach.

1.3 Why are macrocyclic ligands of interest?

It is probably true to say that the study of macrocyclic ligands resulted in a renaissance of inorganic coordination chemistry. This rather bold statement is not intended to imply that macrocyclic ligands *per se* are unique but rather that their appearance occurred at the time in which new theoretical and physical techniques were being developed to allow a better understanding of the structures and reactivity of coordination compounds. Having said that, for much of the past thirty years, coordination compounds of macrocyclic ligands have been the subject of investigation in their own right. The successes of these investigations have changed the way in which we think about coordination chemistry and represent the genesis of supramolecular chemistry and nanochemistry.

Initially, much of the impetus was in the synthesis of macrocyclic ligands and complexes and this lead to the development of highly efficient synthetic routes to large ring compounds. It is still a source of wonder that such large rings can be prepared specifically and in high yield! In later stages, the properties of the macrocyclic complexes themselves became the centre of attention. It was certainly the case that the majority of macrocyclic complexes were both kinetically and thermodynamically more stable than analogous compounds with non-cyclic ligands, and this resulted in an enormous effort, and no little controversy, in trying to understand the origins of these effects.

> Macrocyclic complexes are usually thermodynamically and kinetically more stable than complexes with related non-cyclic ligands.

More recently, the high stability of macrocyclic complexes has been utilized in the construction of models for metalloproteins and in a wide range of technological applications. Particular interest has centred upon the use of macrocyclic ligands for the selective extraction of metals and the development of hydrometallurgical methods to replace conventional, but environmentally damaging, pyrometallurgy.

The study of macrocyclic ligands and their complexes allows us to probe many of the more subtle aspects of the reactivity of coordination compounds which would not be possible in less stable complexes with non-cyclic ligands.

Above all, macrocyclic chemistry is fun! The structures of the molecules are unusual and attractive, very often unexpected results are encountered, and when the synthetic routes have been optimized remarkable large molecules may be prepared in high yields from the most unlikely small molecule precursors.

What are the successes of macrocyclic chemistry? In one sense, the greatest success is in the understanding of macrocyclic chemistry itself! At the beginning of the 1960's, the properties of macrocyclic coordination compounds were genuinely regarded as different and unusual. As the understanding of these compounds developed, so did our view of coordination chemistry in general.

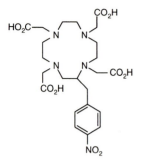

A macrocyclic ligand proposed for the binding of the radioactive nuclide ^{90}Y for use in radiotherapy.

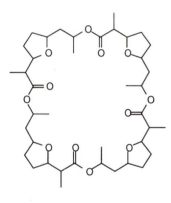

nonactin

Preorganized means that the donor atoms are held in the correct spatial positions for coordination to a metal.

The clinical use of macrocyclic ligands to bind radioactive metals for chemotherapeutic applications or of paramagnetic complexes with lanthanoids as imaging agents is expected to become routine over the next few years.

Numerous microelectronic devices and sensors based upon the use of macrocyclic ligands for the detection, amplification or recognition of metal ions have been proposed or fabricated. In the more sophisticated devices, the binding of a metal ion triggers some mechanical or optical change within the system.

Finally, macrocyclic ligands are not purely a laboratory phenomenon and the development of synthetic macrocyclic chemistry has resulted in an increased understanding of the functions and properties of naturally occurring biological macrocycles.

1.4 Bioinorganic aspects

Macrocyclic ligands and complexes are widespread in biology and it is no exaggeration to state that life as we know it could not exist in the absence of such molecules. Macrocyclic ligands based upon the porphyrin ring system **1.6** and close structural relations such as corrin are ubiquitous. Iron complexes of these ligands are involved in processes as divergent as electron transport, dioxygen transport and dioxygen storage. The various chlorophylls are responsible for the absorption of light and the initiation of electron transfer processes in green plants which ultimately lead to the oxidation of water to dioxygen. Oxygen donor macrocycles such as nonactin are involved in the active transport of alkali metal ions, the control of the ionic balance within cells, the transmission of neural impulses, and represent a class of widely used antibiotics.

One obvious use of simple synthetic macrocycles is to try to make compounds which will behave in a similar manner to the biological molecules, but which possess simpler structures and which may be studied to give a better understanding of the function of the naturally occurring species. Such simple compounds can act as *structural* models which reproduce spectroscopic or other features of the biological molecule or as *functional* models which replicate the biological and chemical reactivity. It is probably fair to say that the success in preparing structural models has been outstanding but that in devising functional model systems has only been moderate.

A more subtle aspect of macrocyclic coordination chemistry lies in trying to model metal-binding sites in proteins which are not formally macrocyclic. A protein can be regarded as a very large, reasonably preorganized, multidentate ligand. The relatively well-defined spatial arrangement of the donor atoms at the metal binding domain could be reproduced by a macrocyclic ligand with equivalent donor atoms arranged in a similar manner. Once again, this approach to making model compounds has been very successful for structural analogues but, as yet, functional models are few and far between.

Current efforts are directed towards the design of macrocyclic systems which are capable of acting as fully functional enzyme mimics and which can catalyse a particular reaction in a highly specific manner.

2 Types of macrocyclic ligand and their complexes

Introduction

This chapter will introduce some of the macrocycles which are commonly encountered in modern coordination chemistry. The list is by no means comprehensive, but should give an overview.

2.1 Polyamine or imine macrocycles

Ammonia and amines have been important ligands in coordination chemistry since the time of Werner. Early in the development of the subject, it was found that complexes with polydentate, chelating ligands were subjectively

The chelate effect states that complexes with polydentate ligands will be more stable than complexes with similar monodentate ligands.

and quantitatively more stable than complexes with monodentate ligands and this phenomenon was described as the *chelate effect*. The effect was found to be sequential and the stability increased as more donors were incorporated into the polydentate ligands. A typical sequence of stabilities is presented in Fig. 2.1.

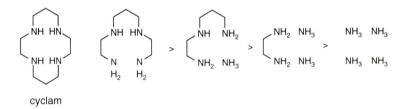

cyclam

Fig. 2.1. The stability of copper(II) complexes containing four nitrogen donors increases with the ligand denticity.

Fig. 2.1 leads to an obvious question, which will be discussed in detail in Chapter 5. Are complexes with cyclam more stable than those with the open-chain tetradentate ligand?

As a result, a vast range of macrocyclic ligands incorporating varying numbers of nitrogen atoms in rings of various sizes have been prepared, although the majority contain four donors (**2.1**). These macrocycles, typified by cyclam, are known as the *tetraazamacrocycle*s. Rings with greater or smaller numbers of donor atoms are less important, with the exception of the small macrocycle tacn which has a rich and interesting chemistry.

tacn

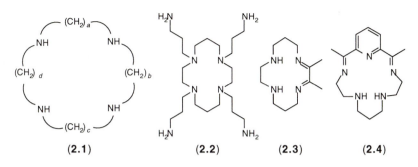

cyclen

(2.1)   (2.2)   (2.3)   (2.4)

Varying the number of methylene spacer groups within the structure **2.1** allows ring size effects to be studied – for example, cyclam may be compared with cyclen. The basic structure **2.1** has been modified in numerous ways. Substituents may be introduced at nitrogen or carbon. Rings may be linked together through spacer groups or fused. Functional substituents may be incorporated to give additional donor sites or sites which may undergo other types of reaction. Such systems are often described as 'pendant arm' macrocycles.

Typically, pendant arm macrocycles have been prepared with some specific application in mind – in other words, the macrocyclic core provides a central structure upon which additional functionality may be built. An example is **2.2** which is readily prepared by the reaction of cyclam with acrylonitrile followed by reduction of the terminal nitrile residue. The exocyclic amino groups are capable of binding to additional metal ions, or to the metal ion within the macrocycle and may also be protonated.

Nitrogen donors are not limited to sp^3-hybridized amines and a vast number of macrocyclic ligands based upon imine or pyridine donors have been prepared and studied. Once again, this represents an extension of the known chemistry of open-chain ligands to cyclic systems. Typical examples include the mixed imine-amine ligand **2.3** and the pyridine macrocycle **2.4**. Once again, numerous highly functionalized derivatives have been prepared.

The ligands may be neutral, anionic or (less commonly) cationic. Anionic ligands may result from deprotonation of NH or other acidic groups and include the fully conjugated systems **2.5** and **2.6**.

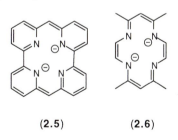

(2.5)   (2.6)

In general, these nitrogen donor macrocycles exhibit a similar coordination behaviour to their open-chain analogues and form complexes which often resemble, but are significantly more stable than, familiar compounds with amines, imines or pyridine donors. This means that the coordination chemistry of these nitrogen donor macrocycles will be dominated by the transition metals, the lanthanoids and the actinoids. One of the enjoyable aspects of macrocyclic ligands is the possibility of imposing unusual coordination numbers or geometries upon metal centres as a result of the preorganized structure. Ligand **2.7** has been designed to impose a pentagonal planar donor set onto a metal centre whilst **2.8** asks what happens when a reasonably flexible ligand presenting seven donor atoms interacts with a metal centre.

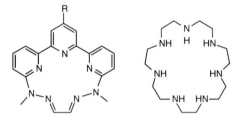

(2.7)   **(2.8)**

2.2 Porphyrins and phthalocyanines

As mentioned in Chapter 1, numerous naturally occurring and biologically important macrocycles are based upon the porphyrin ring system. It is this feature, together with the substantially different properties associated with the rigid, highly preorganized, macrocycles which lead us to treat these compounds separately from other nitrogen donor ligands. A closely related, but synthetic macrocycle is phthalocyanine, which incorporates the four nitrogen donors of the porphyrin, but replaces the *meso* carbon atoms, between the five membered rings, by nitrogen atoms. In addition, each five membered ring has a fused benzo group attached. Although these macrocyclic ligands are just special examples of nitrogen donor ligands, the enormous number of examples and the huge literature justify their special treatment.

The porphyrin ring system is ubiquitous in nature and iron-porphyrin complexes are involved in most aspects of dioxygen binding, transport and metabolism in the higher organisms. Closely related compounds play critical roles in the crucial electron-transfer processes associated with respiration and are involved in many of the most important metabolic oxidation and reduction processes. The iron complexes are associated with proteins in the so-called *heme* (or *haem*) proteins. The iron complex of protoporphyrin IX incorporates many of the structural features which are found in the iron complexes associated with the heme proteins.

The interest in porphyrin derivatives lies in the preparation of model compounds which might mimic the biological species which are involved in electron-transfer and dioxygen-activation processes. The synthesis of simple porphyrins such as *meso*-tetraphenylporphyrin, from benzaldehyde and pyrrole in the presence of Lewis acids is facile. Like porphyrins, phthalocyanines and their complexes are highly coloured, and numerous

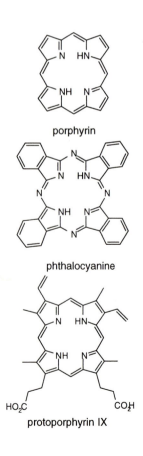

porphyrin

phthalocyanine

protoporphyrin IX

derivatives have found applications as pigments or dyestuffs. Once again, the synthesis of the parent phthalocyanine ring system is facile, and simply heating phthalonitrile with metal salts leads directly to metal complexes.

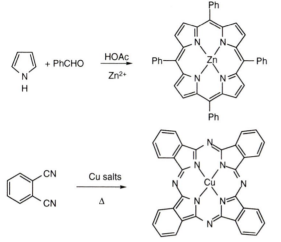

Fig. 2.2. The synthesis of simple porphyrin and phthalocyanine derivatives.

Much synthetic effort has been expended on the preparation of macrocyclic complexes which bind dioxygen in a similar manner to the heme proteins such as hemoglobin or myoglobin. Although cobalt and nickel complexes have been used successfully, the majority of such studies concentrate upon iron compounds and one such model system is seen in **2.9**.

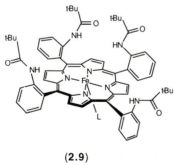

(2.9)
a picket fence porphyrin

One of the major difficulties associated with the modelling of dioxygen binding by iron(II) complexes is the irreversible formation of dinuclear iron(III) μ-oxo bridged species (Fig. 2.3). The picket fence porphyrins were prepared with bulky substituents to prevent such reactions occurring. When the axial ligand L is a substituted imidazole, the picket fence porphyrins are able to bind dioxygen in a reversible manner that closely resembles the behaviour of the natural hemoproteins.

μ-oxo means that an oxygen atom is in a bridging position between two metal centres.

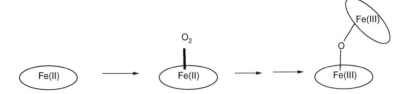

Fig. 2.3. The binding of dioxygen to model iron(II) systems very often leads to the irreversible formation of μ-oxo iron(III) dimers. The use of bulky substituents may prevent this.

corrin

Vitamin B_{12} deficiency results in pernicious anemia.

Other biologically important macrocyclic ligands are based upon the structurally related corrin ring system. The corrin system is a partially reduced derivative of porphyrin, in which one of the *meso* carbon atoms has been replaced by a direct interannular C–C bond. This ring system is found in the Vitamin B_{12} group of cobalt complexes. In Vitamin B_{12} itself the cobalt has a cyanide ligand attached to one of the axial sites and a heterocyclic nitrogen donor of one of the substituents of the corrin in the other. This is shown in Fig. 2.4. The cyanide ligand arises during the isolation of the compound and the biologically active coenzyme contains a C-bonded organic group. The macrocyclic system is capable of stabilizing an organometallic compound in aqueous, physiological conditions!

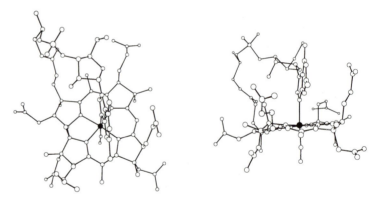

Fig. 2.4. Two views of 2-methyladeninylcyanocobamide (Vitamin B_{12}) as found in the solid state. The first view is looking down upon the corrin ring whilst the second is viewed in the plane of the corrin and shows the axial binding of cyanide and an adenine to the cobalt.

A closely related macrocycle is found in the chlorophylls, which are the magnesium complexes responsible for the harvesting and conversion of light energy to chemical energy in the photosynthetic apparatus of green plants. In these compounds, the porphyrin structure is further elaborated by the fusion of an additional five-membered ring to the periphery of the macrocycle. The long chains are important in the orientation of the molecules within the entire photosynthetic apparatus.

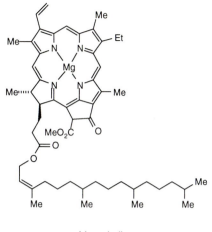

chlorophyll *a*

Porphyrins and phthalocyanines form complexes with most of the elements in the periodic table, regardless of their metallic or non-metallic character. The study of the coordination chemistry of these N_4 donor ligand gives valuable insights into bioinorganic chemistry and also provides some of the more successful examples of biomimetic systems. The structural development of such macrocyclic ligands has exercised the skills and imagination of many synthetic chemists as they try to reproduce the complexities of biological systems under *in vitro* conditions. In part, the interest in these systems stems from their remarkable chemical and thermal stability and derivatives are now finding applications as components in systems designed for molecular electronics and photonics (Fig. 2.5).

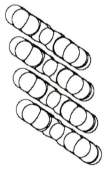

Fig. 2.5. Stacked phthalocyanine molecules in one of the crystal forms of phthalocyanine.

2.3 Crown ethers

The cyclic polyethers (crown ethers) were introduced in Chapter 1 and form a large family of ligands with varying ring sizes and numbers of donor atoms. Ring sizes vary from the 12-membered macrocycle 12-crown-4, **2.10** through 15-crown-5 and 18-crown-6 to compounds with 11 or more oxygen atoms, such as 33-crown-11, **2.11**.

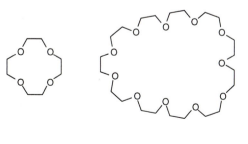

(**2.10**)   (**2.11**)

The crown ethers were an accidental discovery, but it was soon noticed that Group 1 compounds exhibited very strange properties in the presence of the polyethers. For example, when solutions of 18-crown-6 or dibenzo-18-crown-6 in benzene are shaken with solid potassium permanganate, some of the salt dissolves to give a coloured solution ('purple benzene'). In the absence of the crown ether, $KMnO_4$ is completely insoluble in benzene. The metal ion coordination chemistry of the crown ethers is predominantly associated with the elements of Groups 1, 2 and 3, although a wide range of other complexes are known. The structures of the crown ethers, and their coordination behaviour are reminiscent of those of antibiotics such as nonactin and it should be remembered that the cyclic polyethers are potentially extremely toxic and biologically active compounds.

Once again, considerable structural development is possible and numerous derivatives with innocent or functional substituents are known. Crown ethers with fused benzo rings (for example, dibenzo-18-crown-6) are commonly encountered and it is found that the solubility properties vary considerably with substitution. The majority of crown ethers are based upon the $-OCH_2CH_2O-$ motif, but series of compounds are known in which some or all of the subunits are replaced by generic $-O(CH_2)_nO-$ groups.

Although they are still termed crown ethers, many of the larger members of the group are highly flexible and do not necessarily adopt the archetypal 'crown' conformation. A typical example is seen in the severely twisted solid state structure of tetrabenzo(b,e,q,t)-30-crown-10 (Fig. 2.6). Although crown ethers and azamacrocycles have many structural features in common, their coordination behaviour is almost mutually exclusive. In addition, more crown ethers containing large rings and large numbers of donor atoms are known, and have been studied, than azamacrocycles.

Not all crown ethers are crown-shaped!

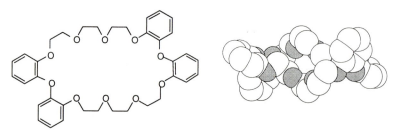

Fig. 2.6. The structure of tetrabenzo(b,e,q,t)-30-crown-10 and the conformation found in the solid state.

As mentioned, structural development is possible and numerous derivatives of crown ethers have been prepared, usually with the aim of modifying the binding of Group 1 or 2 metal ions or with the intention of designing systems whose response to an external stimulus is modified in the presence of these metal ions. A typical example is **2.12** in which the binding of Group 1 metals to the crown ether is likely to modify the photochemical *cis-trans* isomerization of the diazo linker – or alternatively, the binding of the metal ions could be *switched* by the *cis-trans* isomerization.

Molecular switches are important components for molecular electronics. Ligand **2.12** could bind a metal ion in a sandwich between the two crown ethers in the *cis* conformation but not in the *trans*.

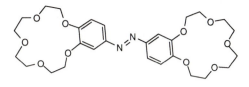

(2.12)

2.4 'Unusual' donor atoms

'Unusual' is an emotive and probably meaningless phrase; in this context, unusual can best be taken to mean donor atoms other than nitrogen and oxygen! Macrocyclic ligands have been designed with most possible donor atoms incorporated into them (Fig. 2.7).

Obvious extensions of the nitrogen and oxygen donors discussed above are the 'softer' analogues with phosphorus, sulfur or selenium donors. Such ligands are known and, although not so extensively studied as their harder congeners, have proved to have a rich chemistry, in particular with the second and third row transition metals.

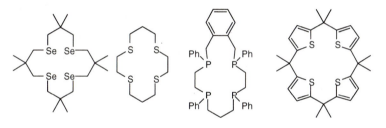

Fig. 2.7. Some macrocyclic ligands with 'unusual' donor atoms.

The widespread use of these macrocyclic systems is limited by the lack of convenient synthetic methods and the use of extremely reactive or toxic intermediates in their synthesis. Notwithstanding the need to use highly toxic $ClCH_2CH_2SCH_2CH_2Cl$ or an analogue for their synthesis, thiacrown ethers, particularly **2.13** and **2.14** have been shown to have a rich coordination chemistry.

(2.13)

(2.14)

2.5 Mixed donor-atom macrocycles

Macrocyclic ligands containing two or more different types of donor atoms are widely used. These ligands have the advantages of combining two or more different bonding preferences allowing new types of selectivity in metal ion binding. Numerous different combinations of donor atoms have been used and a selection of ligands is presented overleaf (**2.15** - **2.18**). Indeed, mixed *N,S*-donor compounds such as **2.15** represent some of the very earliest synthetic macrocyclic ligands. The mixed nitrogen-oxygen donors in ligands **2.16** and **2.17** represent a mixing of the motifs of the azamacrocycles and

2.15 was one of the first synthetic macrocycles to be prepared in a designed *template* reaction.

the crown ethers and were prepared early on in macrocyclic chemistry. The replacement of one of the oxygen donors of a crown ether by nitrogen allows additional functionalization to be achieved in a very facile manner. Compound **2.17** represents another ligand in which a photoactive residue is attached to the macrocycle.

2.16 is a diazacrown and **2.17** is a monoazacrown. This nomenclature may be developed in a non-systematic way for other mixed donor atom ligands. For example, **2.13** could be called tetrathia-12-crown-4 and **2.8** might be called heptaaza-21-crown-7, although the latter has no obvious advantage over [21]aneN$_7$.

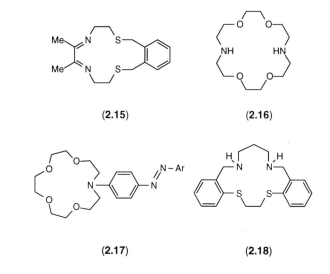

(2.15)　　　　**(2.16)**

(2.17)　　　　**(2.18)**

We will encounter other examples of mixed donor macrocyclic ligands later in this book. Such compounds are of particular interest in maximizing the specificity of a ligand for a particular metal ion and **2.18** represents one of a very large series of ligands prepared to optimize the separation and binding of transition metal ions.

2.6 Cryptands and cavitands and encapsulating ligands

In the same way that macrocyclic ligands are related to open-chain ligands, it is possible to expand the structures into the third dimension to give species which are capable of *encapsulating* a metal ion or other species. Early examples of such ligands were the so-called *cryptands* (which form metal complexes called *cryptates*). This nomenclature arises from the concept of incarcerating a metal ion within the cavity of the ligand. These are extensions of the crown ethers and which make use of bridgehead nitrogen atoms to attach additional functionality, leading to a bicyclic system. The cryptands are named according to the number of oxygen donor atoms which are present in the bridges; thus, **2.19** is known as [2.2.1] and **2.20** as [2.2.2]. Numerous related ligands such as **2.21** which combine nitrogen and oxygen donors have been prepared. The coordination chemistry of cryptands is primarily associated with Group 1 and Group 2 elements although they also have a rich anion coordination chemistry. Compound **2.21** is a curiosity, a 'what if' ligands – what happens if a ligand which binds a Group 1 metal (the diazacrown) is combined with one which binds transition metals (the 2,2'-

(2.19)

(2.20)

(2.21)

bipyridine)? One of the problems with these ligands relates to the role of the bridgehead nitrogen atoms. Although they have been drawn such that the nitrogen lone pair appears to be directed into the macrocyclic cavity, the equilibrium conformation of the free ligand may differ, in which case conformational changes are necessary if the nitrogen atoms are to be involved in binding metal ions.

Although their detailed coordination chemistry is beyond the scope of this book, it is worthwhile introducing a number of miscellaneous oxygen containing ligands at this point. These compounds form complexes with Group 1 and other metals, but interest in them lies more often in the binding and recognition of organic compounds. These molecules represent molecular *hosts* which can bind a variety of *guest* molecules. Early examples were the spherands **2.22** which possess ether oxygen atoms at each end of a tube-shaped molecule (Fig. 2.8); the six oxygen atoms of **2.22** provide an O_6 donor set and lithium and sodium ions are bound very strongly in the centre of the molecule. A related host structure is found in the calixarene family of molecules. These are molecules formed from the condensation of a phenol with formaldehyde in the presence of base and contain from 4 to 8 aromatic rings. A typical example is the calix[6]arene **2.23**. Although the calixarenes certainly interact with Group 1 metal ions, they are more studied because of their ability to incorporate organic guests within the, moderately, hydrophilic cavity. The calixarenes are, like the spherands, cup- or tub-shaped molecules (Fig. 2.9). However, the calixarenes are not conformationally fixed, and a large number of conformations are possible in which the mutual orientation of adjacent rings may be described by 'up-up' or 'up-down' arrangements of the phenolic hydroxy groups leading to a specialized nomenclature (*cone, partial cone, alternate* etc.). These remarkable molecules have a rich host-guest chemistry, and Fig. 2.10 shows the encapsulation of a C_{60} fullerene by two calix[5]arene derivatives.

Host-guest interactions are a very important part of contemporary chemistry. Concepts of *molecular recognition* spring from interactions of this type.

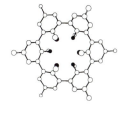

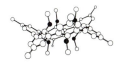

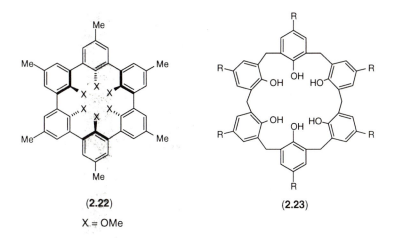

(2.22)

X = OMe

(2.23)

Fig. 2.8. Two views of **2.22** showing the three dimensional structure.

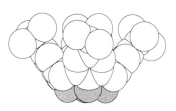

Fig. 2.9. The cup-shaped conformation of tetra(*tert*-butyl)calix[4]arene.

Another series of oxygen containing cavity molecules are the cyclodextrins. These are cyclic oligomers of glucose containing between six and 12 of the sugar units. The molecules are rather rigid and the dimensions of the cavity may be rather precisely defined. Furthermore, the cavity is hydrophobic and guest molecules include numerous non-polar molecules such as arenes and even alkanes. A typical example is γ-cyclodextrin **2.24** which contains eight glucose subunits. Naturally, the presence of the oxygen atoms also means that these ligands also have a rich chemistry with Group 1 and Group 2 metal ions.

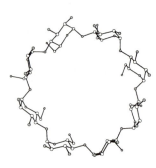

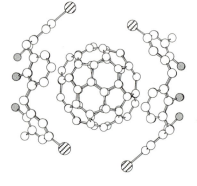

(2.24)

Fig. 2.10. The encapsulation of a fullerene molecule by two calix[5]arene derivatives.

However, the concept of polycyclic systems is not limited to crown ether derivatives and over the years a number of unusual and interesting ligands which can encapsulate metal ions have been described. These include all-nitrogen donors such as **2.25** (sepulchrate), **2.26** and **2.27**, as well as systems with mixed donor atoms, **2.28**. These ligands are commonly obtained directly as transition metal complexes from template reactions (see Chapter 3).

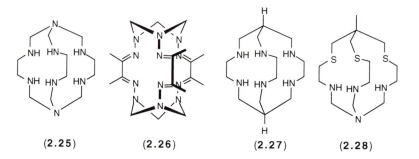

(2.25) **(2.26)** **(2.27)** **(2.28)**

Finally, we can note that the structural development is not confined to the formation of bicyclic ligands; tricyclic systems are of growing interest.

Many of the encapsulating ligands discussed above have one potential problem. The cavity within the host may be ideally suited to the desired guest, but it is not immediately obvious how to get the guest molecule into the cavity! A useful analogy is that the encapsulating ligand acts as a prison. Once the prisoner is in the cell, it is almost impossible to get out without prizing the bars of the cage apart. However, in our case, the gaoler has no key to the cell door and it is also necessary to prize the bars apart to get the prisoner *in*! Some of the consequences will be discussed in Chapters 3 and 5.

2.7 Dinucleating macrocycles and linked macrocyclic ligands

Inorganic chemists are frequently interested in controlling and modifying the interactions between two or more metal centres. Very often, the impetus comes from a desire to understand the functioning of polynuclear sites within metalloproteins or to optimize electron transfer or energy transfer processes in artificial systems.

Two approaches are immediately apparent; two macrocycles may be covalently linked together to so as to bring two bound metal centres into close proximity, or a large macrocyclic ligand may be designed which will bind two metal ions within the same cavity. We have already seen an example of a linked macrocyclic ligand in **2.12**.

Almost any of the macrocyclic ligands that we have considered so far may be linked together, and considerable ingenuity has been applied to the synthesis of such systems over the years. A typical example is **2.29** which may be reduced by diborane to give the parent bis(cyclam) ligand. Molecules such as **2.29**, **2.30** and **2.31** represent unambiguous examples of linked macrocycles in which the two macrocyclic domains are held in such a way that they interact with different metal ions. However, compounds such as *cis*-**2.12** or the more flexible compound **2.32** introduce an ambiguity.

Many metalloproteins contain two or more metal centres in close proximity; hemerythrin contains two Fe centres, hemocyanin two Cu, a range of *iron-sulfur* proteins contain two, three, four or more Fe centres in combination with sulfide ions, the photosynthetic dioxygen evolving centre contains four Mn, laccase contains three Cu *etc.*

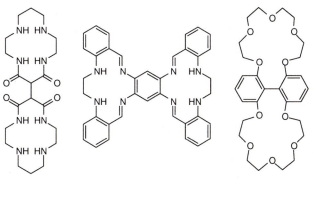

(2.29)   (2.30)   (2.31)

With these ligands, it is possible for both of the macrocyclic rings to form a sandwich with just one metal ion coordinated to both (Fig. 2.11). Even if the structure is sufficiently rigid to prevent this, it is very often found that the coordination of the first metal ion to one of the macrocycles results in sufficient positive charge build -up in the ligand that binding of a second metal ion is very unfavourable or impossible. Nevertheless, numerous dinuclear complexes based upon ligand systems of this type are known.

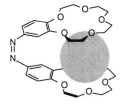

Fig. 2.11. A metal ion sandwiched between the two macrocycles of **2.12**.

Naturally, it is also possible to design ligands such as **2.33**, **2.34** or **2.35** which deliberately form sandwich complexes.

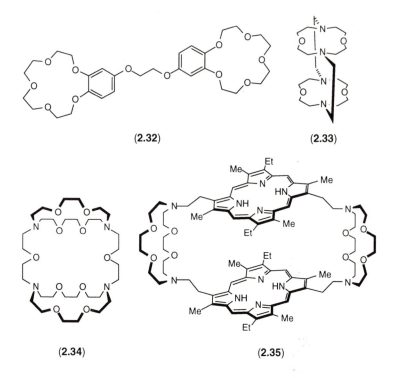

(2.32)   (2.33)

(2.34)   (2.35)

The prefix μ- denotes a bridging ligand

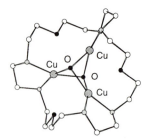

Fig. 2.12. The complex trinuclear cation $[Cu_3(\textbf{2.37})(\mu\text{-OH})_2]^{4+}$

More control over metal-metal interactions may be achieved by the use of larger macrocycles which can bind two or more metal ions, very often with the assistance of an intramolecular or ancillary bridging ligand. Ligand **2.36** contains two spatially separated NS_2 metal-binding domains. Reaction with copper(II) followed by treatment with azide yields $[Cu_2(\textbf{2.36})(\mu\text{-N}_3)_2(N_3)_2]$ in which two of the azide ions bridge the two copper centres and the two remaining azides are bound terminally to each of the metals. The related flexible azacrown ligand **2.37** binds three copper(II) centres which are connected above and below the Cu_3 triangle by two triply-bridging hydroxide ligands in the complex $[Cu_3(\textbf{2.37})(\mu\text{-OH})_2]^{4+}$ (Fig. 2.12).

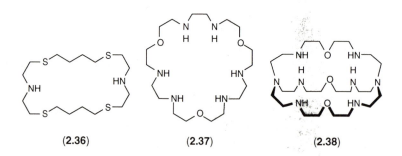

(2.36)   (2.37)   (2.38)

Bicyclic ligands based upon extended cryptands are also capable of forming polynuclear complexes, and **2.38** forms dinuclear complexes with a range of transition metal ions. Of particular interest is the complex $[Co_2(\textbf{2.38})(\mu\text{-}OH)(\mu\text{-}O_2)]^{3+}$ in which the bridging peroxo ligand is formally derived from dioxygen.

Dinucleating macrocycles which can bind two metal ions without the need for additional ligands are also known, and **2.39** provides an example of a bis(phenolic) system which can bind two copper(II) centres. In the complex cation $[Cu_2(\textbf{2.39})]^{2+}$ each metal ion is four coordinate and square-planar and is coordinated to two imine nitrogen donors and two phenolates (Fig. 2.13).

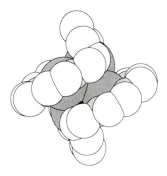

Fig. 2.13. The dinuclear copper complex from **2.39**.

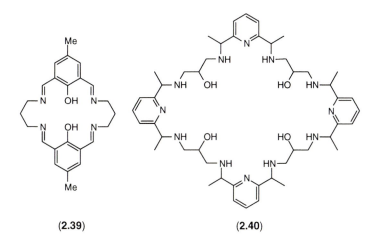

(2.39)   (2.40)

The use of additional donor sites attached to the periphery of the macrocycle allows the assembly of some spectacular complexes, and one notable example is the manganese(II) complex of **2.40**. The complex contains a central Mn_4O_4 cubane core using the four deprotonated hydroxy groups of the ligand (Fig. 2.14). The coordination geometry of each manganese is completed by three Mn-N bonds (Fig. 2.15).

Fig. 2.14. The central part of the tetranuclear manganese complex with **2.40**. The substituents and pyridine rings have been omitted for clarity.

2.7 The concept of hole-size

It will be apparent by now that one of the key features of macrocyclic ligands is the presence of the cavity surrounded by donor atoms. Indeed, our original definition in Chapter 1 is based upon a cavity being sufficiently large to bind a metal ion. In this section we will try to quantify the size of the cavity and look at the consequences of interactions between metal ions and metal-binding cavities which are not the ideal size. The ideas will be developed later in Chapter 5 where the selective binding of metal ions is considered in more detail.

A very simple concept is that of the *hole-size* of a macrocyclic ligand. Unfortunately, it proves to be remarkably difficult to quantify this concept. Intuitively, the cavity within cyclam, with a 14-membered ring is larger than that within cyclen which only has a 12-membered ring. But how can we

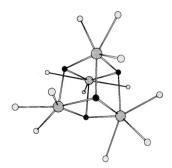

Fig. 2.15. The cubane core showing only the donor atoms.

compare cyclen with 12-crown-4, which also has a 12-membered ring? How can N_4 donor macrocycles such as cyclam be compared with mixed donor atoms systems such as **2.41**? Can the interior cavities of the encapsulating ligands be precisely defined?

The most easily determined parameter is the radius of the cavity, which can then be matched up directly with known metal ion radii in selecting appropriate combinations of metal ion and ligand. The earliest approach was to estimate the radius of the cavity directly by measuring the distance between the nuclei of diametrically opposed donor atoms using molecular or computer-generated models, or by taking distances directly from solid state structural determinations of the free ligands or their complexes.

Using this approach, it is relatively easy to estimate the *hole-size* (radius of the available cavity) for simple ligands such as the tetraazamacrocycles and crown ethers. The radius r that we have measured consists of that of the available cavity $r(H)$ plus the radius of the donor atom $r(D)$. The radii of donor atoms are known (we use Pauling covalent radii here although some workers advocate van der Waals radii) and simple subtraction yields the hole-size. The basic method is presented in Fig. 2.16 for cyclam. By studying a range of cyclam ligands and complexes, r values between 1.90 and 2.10 Å corresponding to $r(H)$ of 1.18 to 1.38 Å are obtained.

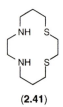

(2.41)

$r(H) = r - r(D)$

Donor	$r(D)$/Å
N (amine)	0.72
N (imine)	0.67
N (pyridine)	0.70
O ether	0.76
P	1.02
S	0.97

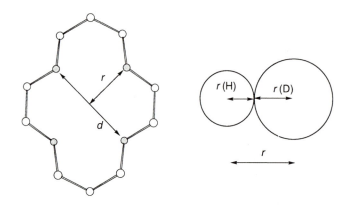

Fig. 2.16. The determination of the macrocyclic hole-size. The distance d between two diametrically opposed donors leads to the radius, r, of the cavity. To obtain the hole-size $r(H)$ it is necessary to subtract the radius of the donor atom $r(D)$.

Why was it necessary to give a range of values for the hole size? One of the principal problems is related to the conformation of the flexible macrocyclic ligand. If the salt $[H_2cyclam][ClO_4]_2$ is studied, two N...N distances of 4.223 and 3.968 Å are observed giving $r(H)$ of 1.395 and 1.260 Å respectively. Which should we take? Two series of nickel(II) complexes with cyclam are known. In the first, the nickel is six-coordinate and octahedral with the cyclam occupying the equatorial sites and two additional axial ligands above and below the cyclam, whereas in the second a low-spin square-planar nickel(II) lies in the macrocyclic cavity and no axial ligands are present. The diameter of the cavity in *trans*-[Ni(cyclam)Cl$_2$] is 4.134 Å ($r(H)$

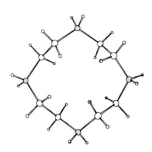

Fig. 2.17. The conformation of cyclen in its hydrochloride salt.

= 1.35 Å) whereas in low-spin [Ni(cyclam)][ZnCl$_4$] it is 3.828 Å (r(H) = 1.19 Å). The flexibility of the ligand allows the nitrogen donors to move to give optimal metal-nitrogen distances. Whilst it is reassuring to have our feeling that low-spin nickel(II) is 'smaller' than the high-spin ion, these measurements bring the first hint that it will only be possible to define an unambiguous hole-size for a completely rigid ligand.

What can we do to compare cyclam with cyclen? If the distance between the nitrogen atoms is measured for the hydrochloride salt of cyclen, a radius of about 2.6 Å (r(H) = 1.9 Å) is obtained, suggesting that cyclen is larger than cyclam! In this hydrochloride salt, cyclen adopts a conformation in which the nitrogen atoms 'point outwards' in an exocyclic mode (Fig. 2.17). Clearly it is necessary to compare similar conformations of the ligands if the comparison is to be meaningful. If the hydrate of cyclen is taken, which has two endocyclic and two exocyclic nitrogen atoms, r(H) values of 1.31 to 1.42 Å are obtained - the same as cyclam! In fact, because cyclen does not form directly analogous complexes to cyclam, it is very difficult to quantify our 'feeling' that cyclen is smaller than cyclam using crystallographic data. The best way to obtain an estimate for cyclen is to draw a best-fit circle through the nitrogen donors of the ligand in a hypothetical conformation in which they are endocyclic and planar.

Exocyclic means that the lone pairs of the donors are oriented away from the central cavity. *Endocyclic* means that the lone pairs are oriented into the cavity.

The simple approach discussed above does allow us to evaluate the effect of substituents. A comparison of *trans*-[Ni(cyclam)Cl$_2$] with *trans*-[Ni(**2.42**)F$_2$] indicates an increase in hole-size from 1.34 to 1.41 Å upon introduction of the methyl substituents, an effect that may be traced to steric interactions between the substituents and the axial ligands.

The crown ethers present a much wider series of ligands which allow us to probe hole-size effects in more detail. It is necessary to modify the method of determination, because ligands such as 15-crown-5 do not have any diametrically opposed oxygen atoms. It is better to define the hole-size as the radius of the best fit circle through the donor atoms. However, even this method has its problems if the conformation of the free ligand or of the ligand in a particular complex is markedly non-planar. Our ideas of hole size are really only valid for a planar or near-planar set of donor atoms. Fortunately, many simple crown ethers do form complexes with the donor sets approaching planarity. Once again, we have to take into account the conformation of the ligands in the complex, rather than that in the free ligands. Some typical values are given in Table 2.1. The uncertainties are great and it is clearly not a useful concept to talk about a *unique* hole-size for a particular ligand.

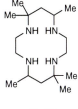

(2.42)

Table 2.1. Hole-sizes of crown ethers and related ligands (calculated from solid state analyses).

Ligand	r(H)/Å
cyclam	1.2-1.4
12-crown-4	1.3-1.5
15-crown-5	1.5-1.9
18-crown-6	1.8-2.2

Clearly the presence of mixed donor atoms will further complicate the calculation of the hole-size, as will significant deviations from planarity or the introduction of unsaturated bonds into the ring structure. With mixed donor macrocycles it will be necessary to take into account the varying covalent radii of the donor atoms. The introduction of unsaturation will normally be beneficial, as it reduces the number of conformations which are available. Furthermore, any additional ligands or anions coordinated to the metal ion are likely to perturb the metal-ligand distances and geometry. Various attempts have been made to more fully quantify the concept of hole-size with varying degerees of success. We will now take the *concept* and see what the consequences for coordination chemistry are, without going into details of these calculations. Further details for the interested reader are given in the suggestions for further reading at the end of the book.

2.8 Some consequences of the hole size

As discussed above, a given ligand has a loosely defined cavity available for binding guest ions or molecules. We might expect to find the best binding of the guest if its fit into the macrocyclic cavity is optimal. In terms of metal coordination chemistry, this refers to the ability to form metal-ligand bonds with the optimal bond lengths and with the distribution of the donor atoms in a favourable geometrical arrangement about the metal. Furthermore, these optimal metal-ligand interactions should be achieved without the introduction of unfavourable steric interactions within the ligand framework.

Ion	r(covalent)/Å
Li^+	1.34
Na^+	1.54
K^+	1.96
Rb^+	≈2.00
Cs^+	≈2.20
Mg^{2+}	=1.45
Ca^{2+}	≈1.60
Sr^{2+}	≈1.80
Ba^{2+}	≈2.00

What exactly does $r(H)$ refer to and what does it tell us? As we decided to use covalent radii for the donor atoms it is appropriate to choose covalent radii for the metal centres, even though these may not be the intiuitive values to use for complexes containing metal ions! However, equivalent results are obtained if we choose to use van der Waals radii for the donors and ionic radii for the metal ions. We commence our discussion with crown ethers and then extend the observations to macrocyclic ligands with other donor sets.

As mentioned previously, the crown ethers form complexes primarily with alkali and alkaline earth metal ions. We can now try to match the sizes of the crown ethers to Group 1 and Group 2 metal ions. The smallest crown ether, 12-crown-4 is more-or-less compatible with the lithium cation and a series of 1:1 complexes with lithium salts are known. However, the lithium ion is a little too large for the cavity and typically lies somewhat above the mean plane of the four oxygen donors. In the salts [Li(12-crown-4)X] (X = Cl or SCN) the anion is also coordinated to the lithium giving a five-coordinate square-based pyramidal geometry. It is rather like the argument about the chicken and the egg to ask if the lithium rises out of the plane because of the axial ligand or whether the axial ligand binds because the metal ion sits above the macrocycle. A similar effect is seen with 15-crown-5 and Na^+, and the sodium cation lies somewhat above the five oxygen donors in the complex [Na(benzo-15-crown-5)(H_2O)]I. The lithium lies about 0.93 Å above the O_4 least-squares plane in [Li(12-crown-4)Cl] whilst the sodium is

0.75Å above the O_5 least-squares plane of the macrocycle in [Na(benzo-15-crown-5)(H$_2$O)]I (Fig. 2.18).

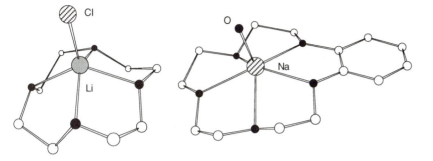

Fig. 2.18. Two complexes in which the metal ion is a little too large for the macrocyclic cavity – [Li(12-crown-4)Cl] and [Na(benzo-15-crown-5)(H$_2$O)]I.

The fit of the potassium ion into 18-crown-6 and related macrocycles is considerably better, and the metal lies only 0.1 Å above the least-squares O_6 plane in [K(dibenzo-18-crown-6)I] (Fig. 2.19). Once again, an axially bound anion is present.

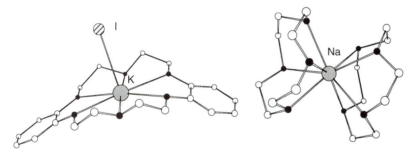

Fig. 2.19. The fit of the potassium ion in the macrocyclic cavity in [K(dibenzo-18-crown-6)I] is good, whilst the sodium cation is too large for the 12-crown-4 ligand and a sandwich species is present in [Na(12-crown-4)$_2$][ClO$_4$].

When the metal ion is a little too large for the macrocyclic cavity, we expect to see it sitting above the donor set, as indeed we do in the lithium and sodium complexes above where the fit is not perfect. An extension of this pattern leads to a second structural motif, in which two of the macrocyclic ligands sandwich the metal ion. This type of complex is found in the complex [Na(12-crown-4)$_2$][ClO$_4$] in which the average O-Na bond length is 2.5Å and the sodium lies close to the mid point (1.5Å) between the two O_4 best planes (Fig. 2.19). Similar arguments lead to the prediction of sandwich complexes between potassium ions and 15-crown-5 and this indeed occurs in complexes such as [K(15-crown-5)$_2$]I.

However, lest we become overconfident, a few cautionary words are necessary. Although Li$^+$ ions are approximately the right size for 12-crown-4, and we have seen above that 1:1 complexes may be isolated, a second

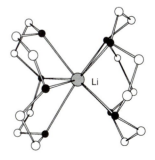

Fig. 2.20. The [Li(12-crown-4)$_2$]$^+$ cation in [Li(12-crown-4)$_2$][Ph$_2$As].

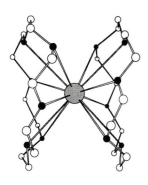

Fig. 2.21. The cation in [Cs(18-crown-6)$_2$]Na.

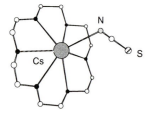

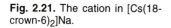

Fig. 2.22. The 1:1 complex [Cs(18-crown-6)(NCS)].

series of Li:12-crown-4 complexes may be isolated. These contain sandwich [Li(12-crown-4)$_2$]$^+$ cations (Fig. 2.20). It is not even completely correct to state that the sandwich structures result when the metal ion is too large for the macrocyclic cavity. Even though Cs$^+$ cations are significantly larger than the cavity in 18-crown-6, it is possible to obtain complexes which contain the [Cs(18-crown-6)$_2$]$^+$ sandwich cation (Fig. 2.21). An example is seen in the compound [Cs(18-crown-6)$_2$]Na which contains the rather unusual sodide Na$^-$ anion; the function of the cation is to stabilize this rather reactive anion. However, it is also possible to obtain a 1:1 complex of CsSCN with 18-crown-6 in which the caesium lies slightly above the best plane of the six oxygen donors and is also coordinated to the nitrogen of the thiocyanate (Fig. 2.22). It is becoming clear that the structure and stoichiometry of the isolated solid state species is very dependent upon the stoichiometry of the reaction and the presence of any additional anions which may coordinate to the metal ion. As always, it should be mentioned at this point that the isolation of a particular solid state species says nothing *a priori* about any solution species which might be present, nor indeed about the relative stabilities of any solution species.

What happens if the metal ion is too small for the bonding cavity of the macrocycle? Without any need to quantify the hole size, if Cs$^+$ is too large for 18-crown-6 and K$^+$ is the correct size, then it is likely that Na$^+$ will be too small for the bonding cavity. A number of possible things could happen. Only some of the donor atoms might be bonded to the metal ion, or in the case of a dramatic mismatch in the hole-size and the size of the cation, two or more metal ions might be incorporated. Another possible consequence is that the macrocyclic ligand undergoes a radical conformational change to adopt some sort of a folded conformation which optimises the metal-donor atom distances. A series of complexes containing {Na(18-crown-6)} cations are known, but the structures are very variable (Fig. 2.23). In the cases of a salt with a porphyrinatoiron(III) complex anion the very distorted structure shown in Fig. 2.23 a is found with Na-O bond lengths varying from 2.364 Å to 2.800 Å, whereas in [Na(18-crown-6)(H$_2$O)](SCN) the distorted cation shown in Fig. 2.23 b is formed in which the folding of the ligand generates six approximately equal Na-O bonds. In contrast, the change of the anion to a ruthenium cluster leads to the cation shown in Fig. 2.23 c with five short and one long Na-O bonds. This latter structure approximates to what we might expect, with one of the oxygen atoms 'folding' over to encompass the sodium cation.

However, the structures are extremely dependent upon the anion and any other ligands which might bond to the sodium and in the cation [Na(18-crown-6)(thf)$_2$]$^+$ the Na-O bonds to the crown ether are more or less equivalent and the crown ether adopts the near planar arrangement found in the potassium complex (Fig. 2.24).

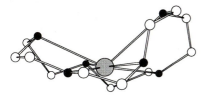

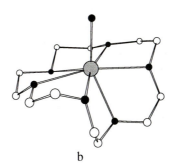

a

b

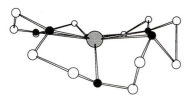

c

Fig. 2.23. Three different [Na(18-crown-6)]$^+$ cations found in salts with different anions; a) in this cation the Na-O bonds are variable and the structure is very distorted whereas in b) and c) five Na-O bonds are similar and one long Na-O bond completes the coordination sphere.

With the smaller lithium cation, only two of the crown ether oxygen donors are coordinated in the complex [Li(18-crown-6)(H$_2$O)$_2$][ClO$_4$] (Fig. 2.25 a).

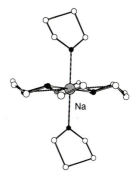

Fig 2.24. The [Na(18-crown-6)(thf)$_2$]$^+$ cation showing the regular conformation of the 18-crown-6 ligand.

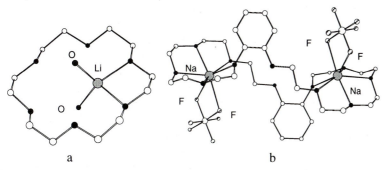

a

b

Fig 2.25. The cations present in a) [Li(18-crown-6)(H$_2$O)$_2$][ClO$_4$] and b) [Na$_2$(dibenzo-36-crown-12)(PF$_6$)$_2$].

When the mismatch in the sizes of the cation and the macrocyclic hole is even larger, or the number of available donor atoms is too great we expect to find even more distorted structures or polynuclear complexes. The large crown ether dibenzo-36-crown-12 forms a dinuclear adduct with NaPF$_6$ in which each sodium is coordinated to five of the crown ether oxygen donors and two fluorine atoms of the 'innocent' hexafluorophosphate anion (Fig. 2.25 b). The somewhat smaller ligand dibenzo-30-crown-10 wraps around the metal ion in the complex [K(dibenzo-30-crown-10)]I in a manner that resembles the seam of a tennis ball (Fig. 2.26). Once again, there is no well defined trend in the structures, and dibenzo-24-crown-8 forms a dinuclear adduct with KSCN in which the two potassium ions are bridged by *N*-bonded thiocyanate ligands (Fig. 2.27).

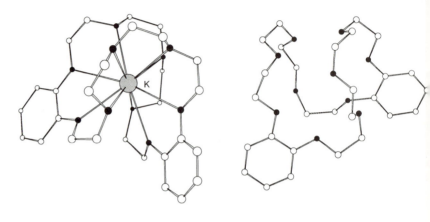

a   b

Fig. 2.26. a) The [K(dibenzo-30-crown-10)] cation and b) the conformation of the ligand in [K(dibenzo-30-crown-10)]I.

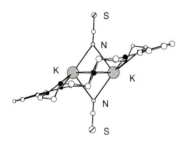

Fig 2.27. The structure of [K$_2$(dibenzo-24-crown-8)(SCN)$_2$].

The situation with cryptands is a little easier, although even here it is not possible to unambiguously define a hole-size. Cryptand ligands may exhibit various conformations depending upon the distribution of the nitrogen lone pairs. They could both be exocyclic, pointing away from the cavity, giving the *exo, exo* conformation, or both could be endocyclic in the *endo, endo* conformation or one could be exocyclic and one endocyclic in the *endo, exo* conformation (Fig. 2.28.). The free ligand [2.2.2] adopts the *endo, endo* conformation in the solid state (Fig. 2.29).

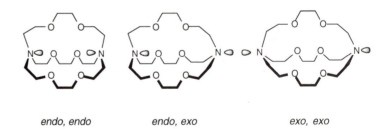

endo, endo   *endo, exo*   *exo, exo*

Fig. 2.28. The three conformations of the cryptand [2.2.2].

Upon coordination to a metal ion, any one of these conformations could, in principle, be involved. Intuitively, one might expect the *endo, endo* conformation might be the most favoured since it has the possibility of giving additional N-M interactions which might stabilize the complex. However, this stabilization is expected to be dependent upon how well the metal ion is matched to the cavity of the cryptand.

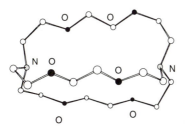

Fig. 2.29. The *endo,endo* conformation of the free ligand [2.2.2] in the solid state.

In fact, the cryptand [2.2.2] forms 1:1 complexes with Na⁺, K⁺, Rb⁺ and Cs⁺ cations within the cavity, each of which exhibits the *endo, endo* conformation (Fig. 2.30). How can this be? In the case of the potassium complex, the K-O distances are in the range 2.776 - 2.791 Å and the nitrogen atoms have a significant interaction with the potassium ions (K-N, 2.874 Å). The binding of the non-optimal metal ions, sodium, rubidium and caesium, is achieved by the introduction of significant amounts of strain within the ligand structure. This is illustrated in Fig. 2.31 which shows the conformation of the ligand in the complexes [K([2.2.2])]I and [Cs([2.2.2])](SCN). The strain may be analysed, *inter alia*, in terms of the relative angles which the two triangles defined by the oxygen donor atoms at each end of the ligand make with one another. A consequence of this twisting of the two O_3 donor sets is the elongation of the cryptand; in [K([2.2.2])]I the non-bonded N...N distance along the long axis of the cation is 5.748 Å whereas in [Cs([2.2.2])](SCN), the twisting lengthens this to an N...N distance of 6.095 Å.

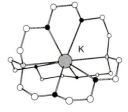

Fig. 2.30. The [K([2.2.2])]⁺ cation.

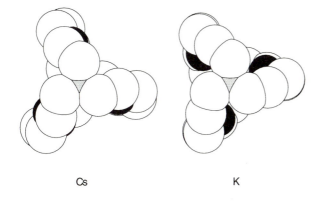

Cs   K

Fig. 2.31. The conformation of the [2.2.2] ligand in the complexes [K([2.2.2])]I and [Cs([2.2.2])](SCN). The metal ions have been omitted for clarity and the ligand is aligned with the N...N axis perpendicular to the plane of the paper.

Similar effects are observed with the tetraazamacrocyclic ligands. The hole-size of cyclam is more-or-less ideal for nickel(II) ions. However, both high-spin, octahedral and low-spin, square-planar nickel(II) ions are a good fit for the cyclam cavity. This is in accord with the experimental observations that both four-coordinate and six-coordinate nickel(II) complexes are formed. The ligand is sufficiently flexible that small conformational changes allow the optimisation of the nickel-nitrogen distances. Thus, in square-planar [Ni(cyclam)]I$_2$ (Fig. 2.32 a) in which the iodide anions are not coordinated to the metal, the Ni-N distances lie between 1.940 and 1.959 Å whereas in the octahedral complex *trans*-[Ni(cyclam)Cl$_2$] in which the two chloride ligands occupy the axial sites the Ni-N distances are between 2.066 and 2.067 Å (Fig. 2.32 b).

These differences are readily explained in terms of the d^8 electronic configuration of the metal ions; in square-planar, low-spin nickel(II), two electrons are placed in the d_{z^2} orbital, whereas in the high-spin octahedral ion, one electron is placed in each of the d_{z^2} and $d_{x^2-y^2}$ orbitals. In the high-spin case, the electron in the $d_{x^2-y^2}$ orbital is oriented directly towards the nitrogen donors of the cyclam. The electron-nitrogen lone pair repulsion results in longer Ni-N bonds and leads to the statement that the high-spin nickel(II) ion is 'larger' than the low-spin ion.

Ligand field effects result in square-planar low-spin nickel(II) ions appearing to be smaller than high-spin octahedral nickel(II).

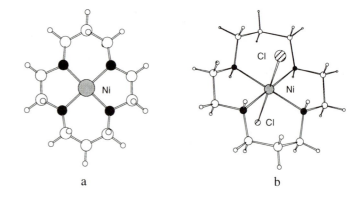

a   b

Fig. 2.32. The structure of a) the square-planar low-spin cation [Ni(cyclam)]$^{2+}$ and b) high-spin octahedral *trans*-[Ni(cyclam)Cl$_2$]. In a), the hydrogen atoms lie below the plane of the paper.

The smaller macrocycle cyclen forms 1:1 high-spin complexes with NiX$_2$ (X = various anions) which contain octahedral cations such as [Ni(cyclen)(H$_2$O)$_2$]$^{2+}$. However, in these complexes, the macrocyclic ligand adopts a folded conformation such that the two additional donor atoms are arranged in *cis* positions. The structure of the [Ni(cyclen)(H$_2$O)$_2$]$^{2+}$ cation found in [Ni(cyclen)(H$_2$O)$_2$](ClO$_4$)$_2$ is illustrated in Fig. 2.33. The Ni-N bonds vary between 2.07 and 2.10 Å and are very similar to the bond lengths observed in cyclam complexes containing high-spin nickel(II). This illustrates once again the way in which a ligand may change conformation in order to optimise metal-ligand bonding distances.

Before finishing the discussion of cyclam and related ligands, it is worthwhile making some additional comments on the conformation of the ligands. Although free amines usually have very low inversion energies, coordination to a metal ion considerably retards these processes. In the case of

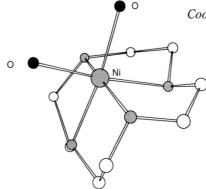

Fig. 2.33. The cation present in [Ni(cyclen)(H$_2$O)$_2$](ClO$_4$)$_2$ showing the folded macrocyclic ligand and the *cis* arrangment of the water ligands.

a cyclam ligand, coordination of the four nitrogen donors to a metal ion generates four new sp^3 centres.

Each of these new sp^3 sites is a chiral centre and a metal complex of cyclam could exist as seven diastereomers. The coordination of the ligand to the metal sufficiently hinders the inversion process that some of the individual diastereomers may be isolated. The diastereomers are best visualized in terms of the relative orientation of the N-H bonds as shown in Fig. 2.34. The majority of cyclam complexes in which the ligand occupies the equatorial plane exhibit conformation III.

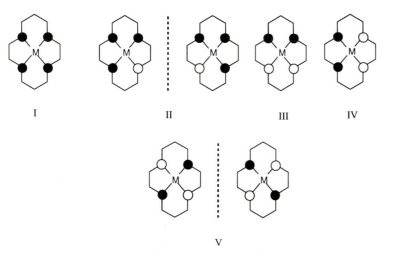

Fig. 2.34. The possible diastereomers of a metal-cyclam complex. A solid circle represents a hydrogen atom above the plane of the paper and an empty circle, one oriented below the plane of the paper. Conformations II and V exist as a pair of enantiomers. Conformer III is the one which is usually encountered.

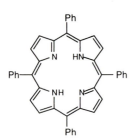

meso-tetraphenylporphyrin

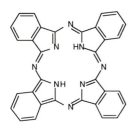

phthalocyanine

pyridine

Finally, we consider the consequences of hole-size mismatch with rigid and relatively inflexible ligands such as porphyrins and phthalocyanines. The extensive delocalization in the ligands means that a planar conformation is favoured on electronic grounds. The basic donor set consists of four nitrogen atoms within a 16-membered ring for both ligands. Our usual hole-size analysis leads to values of $r(H)$ in the range 1.1 - 1.35 Å. This suggests that the cavity is approximately the correct size (although perhaps a little small) for first row transition metal divalent cations or Li^+. Coordination of porphyrins or phthalocyanines to metal ions normally involves the doubly deprotonated dianionic forms of the ligands.

Compounds such as $\{Li_2(pc)\}$ (H_2pc = phthalocyanine) must formally contain two lithium cations and a pc^{2-} anion; however, there is not sufficient room for two lithium ions within the macrocyclic cavity and the compound should be formulated as $Li[Li(pc)]$. The $[Li(pc)]^-$ anion is known in a number of salts and is essentially planar (Fig. 2.35 a). In contrast, the complex $[Ni(tpp)]$ (H_2tpp = *meso*-tetraphenylporphyrin) is decidedly concave, presumably as a result of a mismatch between the cavity size and the metal ion (Fig. 2.35 b). This sort of deviation is to be expected if the metal ion is a little too small for the cavity. Support for this comes from the observation that larger, second row metal cations, give rise to essentially planar complexes, as seen, for example, in $[Ru(tpp)(CO)(EtOH)]$. However, this analysis is certainly over-naive, as the octahedral complex $[Ni(tpp)(py)_2]$ (py = pyridine) contains a near-planar nickel-porphyrin core.

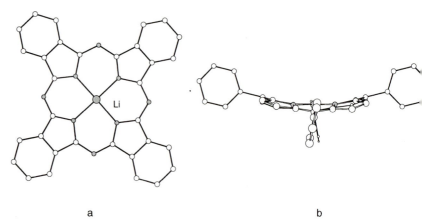

a   b

Fig. 2.35. a) The planar $[Li(pc)]^-$ anion and b) the concave $[Ni(tpp)]$.

When the mismatch in sizes is more dramatic, similar effects to those observed with crown ethers are observed. For example, in the complex $[U(tpp)Cl_2(thf)]$ (thf = tetrahydrofuran), the uranium is coordinated to the two chloride ions and the thf molecule and lies about 1.29 Å above the plane of the four nitrogen donors (Fig. 2.36).

It was mentioned earlier that triazacyclonane has a rich chemistry, and this is dominated by the formation of complexes in which metal ions sit above the plane of the three nitrogen donors. The interest in this ligand lies in the observation that the three donor atoms are constrained to occupy three *facial*

tetrahydrofuran

sites in octahedral complexes and an analogy has been drawn with 'organometallic' complexes containing cyclopentadienyl ligands. Numerous polynuclear complexes have been prepared and have been proposed as structural and/or functional models for metalloenzymes.

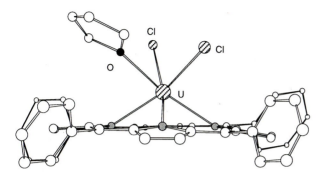

Fig. 2.36. The structure of [U(tpp)Cl$_2$(thf)], in which the metal lies above the plane defined by the nitrogen donor atoms of the macrocycle.

Our survey of metal complexes has been brief, but it should have left the reader with the impression that macrocyclic ligands have a rich coordination chemistry and that, in order to optimize metal-ligand interactions, a variety of bonding modes may be adopted. In the following chapters, we will see how macrocyclic ligands are prepared and discuss some additional consequences of hole-size mismatches.

3 Synthetic aspects 1. The free ligands

Introduction

This short chapter will introduce some of the methods which are used for the synthesis of macrocyclic ligands. It does not intend to provide a rigorous introduction to the organic synthetic methods employed but will merely give an overview of the strategies adopted.

3.1 Non-templated syntheses of free ligands

The meaning of the expression 'non-templated' will become clear in the next chapter. For the time being, we note that it refers to the traditional approach to the synthesis of coordination compounds in which the final step of the reaction involves the reaction of a pre-formed organic ligand with an appropriate metal source. It is a pre-requisite of this approach that the organic ligand must be stable and isolable and the first step in much contemporary coordination chemistry involves time-consuming ligand synthesis.

The majority of macrocyclic ligands of interest to the coordination chemist may be categorized as having *medium* or *large* rings. Although organic synthetic methods for the preparation of *small* and *normal* sized rings are well-developed and highly specific, the same cannot be said for routes to medium and large rings. What are the problems associated with the synthesis of such compounds?

The first factor to consider is the ring strain. This arises when the cyclic structure requires non-optimal bond angles and bond lengths and when a conformation is forced upon the ring such that unfavourable steric interactions occur between substituents. For example, a three membered ring will have internal angles close to 60°, which is considerable smaller than the angle typically associated with sp^3 hybridized carbon atoms. For rings containing 12 or more atoms, the structures are sufficiently flexible that ideal bond lengths and angles may be obtained in strain-free conformations and the ring strain is essentially zero.

The dominant factor controlling the synthesis of large ring compounds is entropic. Although large ring compounds may be prepared, in principle, from any number of components, in the majority of cases, the final reaction step involves the cyclization process in which two ends of a chain, bearing mutually reactive functionalities, come together to create the ring-forming bond (Fig. 3.1). Even if the stoichiometry of the reaction involves two or more components, unless concerted reactions are involved, which is not

Ring size

3,4	*small*
5–7	*normal*
8–11	*medium*
≥ 12	*large*

Macrocyclic ligands of interest to coordination chemists are essentially strain free.

usually the case with large rings, a sequential reaction pathway means that the cyclization step will resemble the process presented in Fig. 3.1.

Fig. 3.1. A schematic view of the cyclization step involved in macrocycle synthesis. The filled and open circles represent mutually reactive functional groups. The first step involves the adoption of a conformation in which the two reactive groups are close to one another and in the correct orientation for reaction. Finally, the new bond is formed and the macrocycle is formed.

The problem is illustrated in Fig. 3.1 – before the cyclization reaction can take place, the two reactive groups must be brought close to one another. Unless there are special secondary interactions within the chain, or unless highly preorganized or rigid systems are involved, the extended conformation is more likely than that which leads to reaction. In other words, it is relatively unlikely that a reactive group will meet the reaction partner at the other end of its own molecule – it is far more likely that it will meet a reactive functionality of a second molecule. The results of this entropic constraint is that the formation of polymeric species will be a significant alternative pathway to the desired macrocyclization reaction.

One widely adopted method that avoids these problems involves the use of '*high dilution*' reaction conditions. The probability of a molecule meeting a second molecule decreases as the concentration decreases; therefore, the more dilute the solution, the better the ratio of macrocycle to polymer. Typical high dilution reactions involve milligrams or tens of milligrams of reactants in solvent volumes of the order of litres! It is also usually necessary to ensure that the reactants are mixed very slowly and special apparatus is required which allows very slow, consecutive, addition of reactants to the reaction solvent over hours or days. Many of the reactive precursors are water sensitive and it is essential to work under strictly anhydrous reaction conditions.

A high dilution reaction with 0.1 mmol of a hydrolytically unstable compound in one litre of solvent would only require 1.8 mg of water to completely destroy the reactant – that is only 0.0002% (w/v) of water in the solvent!

3.2 The ring-formation reactions

By definition, the macrocyclic ligands contain donor atoms. These are very often atoms such as nitrogen, oxygen, phosphorus or sulfur which are either more electronegative than carbon or which may readily be converted to anions which can act as nucleophiles. Accordingly, much of the synthetic methodology for the preparation of free ligands is based upon intramolecular or intermolecular nucleophile-electrophile reactions.

A typical example is seen in the formation of some mixed donor atom sulfur-arsenic macrocycles. The key reactive functionalities are electrophilic CH_2Hal and nucleophilic thiolate groups. The reaction of **3.1** with sodium sulfide under high dilution conditions yields the As_2S macrocycle **3.2** whilst reaction with $HS(CH_2)_3SH$ and KOH gives the As_2S_2 species **3.3**. In these cases, the yields of the macrocyclic ligands are respectable (25-45%), although this is not always true for high dilution reactions. Similar reactions are used for the preparation of thiacrown ethers and related mixed donor atom ligands containing sulfur.

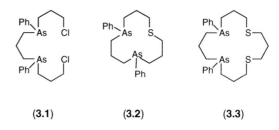

(3.1)   (3.2)   (3.3)

[1+1] means that the macrocycle is derived from one molecule each of two reactants.

[2+2] means that the macrocycle is derived from two molecules each of two reactants.

A hint that the reactions are not quite so simple as indicated is seen in the supposedly complementary reaction sequences **1** and **2** in Fig. 3.2. These reactions give very different ratios of the [1+1] and [2+2] macrocycles. The reaction conditions involve treatment with sodium or potassium hydroxide to deprotonate the thiols, and this is a classical case where an apparently simple reaction may in fact be exhibiting a sodium-template effect. We will say no more at this point.

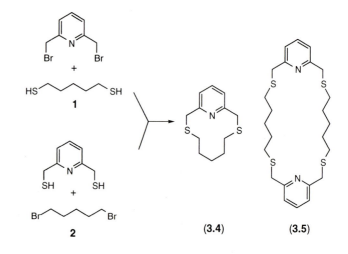

(3.4)   (3.5)

Fig. 3.2. Two complementary cyclization pathways which give rather different ratios of the [1+1] and [2+2] macrocyclic ligands.

The preparation of cryptands is usually performed under high dilution conditions. The final few steps of a typical synthesis of [2.2.2] are presented in Fig. 3.3. In this case, the electrophile of choice is not an alkyl halide but an acid chloride. The reaction of the acid chloride with the amine groups of the diazacrown gives an intermediate amide that is subsequently reduced by BH_3.thf. The cryptand is obtained in about 50% yield and the synthetic method may be extended to the preparation of cryptands with various numbers of oxygen donors within the bridges.

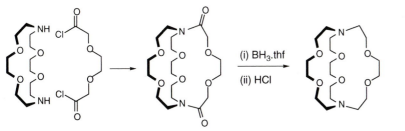

Fig. 3.3. The synthesis of cryptands involves an intermediate diamide that is reduced with a borane derivative.

High dilution reactions are also used for the synthesis of specifically functionalized porphyrin derivatives. Key reaction steps are of the type shown in Fig. 3.4, which illustrates the synthesis of a diaryl-substituted ligand.

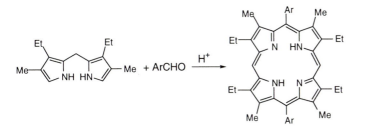

Fig. 3.4. A high dilution reaction is used in the synthesis of specifically functionalized porphyrin derivatives. Typically, the reaction will be performed in CH_2Cl_2 with trifluoroacetic acid.

3.3 When high dilution is not necessary

In some cases, it is not necessary to use high dilution reactions for the synthesis of macrocyclic ligands. If rigid subunits are to be incorporated, the various acyclic intermediates may be pre-organized in the correct conformation for cyclization. Sometimes these reactions are quite spectacular. An early example is seen in the reaction of pyrrole with ketones in the presence of acid, which gives excellent yields of the macrocycle **3.6** in a [4+4] cyclization reaction. Similar reactions are observed with furans to give **3.7**; the reactions are sometimes so favoured that they are explosive.

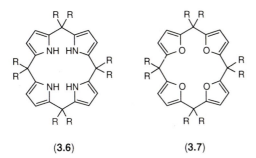

(3.6)   (3.7)

The tosylate group 4-MeC$_6$H$_4$SO$_2$ serves multiple roles in these reactions. Firstly, it is used to protect the secondary amine of **3.8** and central amine of TsO(CH$_2$)$_2$NTs(CH$_2$)$_2$OTs. Secondly, it provides the leaving groups which are lost from TsO(CH$_2$)$_2$NTs(CH$_2$)$_2$OTs and thirdly it activates the terminal NH groups of **3.8** such that deprotonation is possible under moderate conditions.

Surprisingly, the usual routes to tetraaza macrocyclic ligands also involve reactions which do not require high dilution conditions, even though rigid substructures are not present. In the synthesis of the tetraazamacrocycle **3.10**, the key intermediate is the tri-*N*-tosylated species **3.8**. The two terminal NH groups are acidic as a result of the ability of the electron-withdrawing tosyl group to delocalize the negative charge of the anion. The reaction of the dianion with the tosylate TsO(CH$_2$)$_3$NTs(CH$_2$)$_3$OTs occurs under normal concentration conditions to give the tetra-*N*-tosylated compound **3.9** in good yield. Deprotection of **3.9** by stirring in concentrated sulfuric acid gives **3.10** in good yield (Fig. 3.5). Similar reactions may be used to prepare other tetraaza macrocyclic ligands as well as macrocycles incorporating more than four nitrogen donors. Exactly why these reactions proceed under normal concentration conditions is not fully clear, but it seems likely that the bulky tosylate groups play some role in pre-organizing the open chain precursors. As always, the involvement of strong bases which introduce alkali metal cations into the reaction mixture leaves the possibility of template effects open.

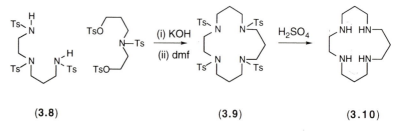

(3.8)   (3.9)   (3.10)

Fig. 3.5. The synthesis of aza macrocycles often involves tosylated species.

en = H$_2$NCH$_2$CH$_2$NH$_2$

Secondary interactions may be important in controlling the course of a cyclization reaction and may allow otherwise unexpected reactions to proceed successfully at 'normal' concentrations. The reaction of [Hen]$^+$ salts with acetone gives respectable yields of **3.12**. The detailed mechanism is not certain, but it is likely that the cyclization step involves an intermediate species such as **3.11** in which hydrogen-bonding stabilizes the required conformation (Fig. 3.6).

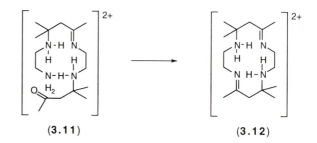

(3.11)   (3.12)

Fig. 3.6. Hydrogen-bonding plays a role in the formation of the Curtis macrocycle. Intramolecular hydrogen bonds stabilize the intermediate and the product.

This macrocycle is known as the Curtis macrocycle after its discoverer and will be encountered again in the next chapter. This macrocycle is also of significance in that it introduces another type of reaction which is of particular importance in macrocyclic chemistry – the formation of an imine from an amine and a carbonyl compound. Imines are good π-acceptor ligands and imine macrocycles are extremely important in stabilizing unusual oxidation states. Imine macrocycles are very often prepared using *template reactions* (see Chapter 4).

Imines contain the $R_2C=NR$ group and are formed in a condensation reaction of an amine with a carbonyl compound.

3.4 The crown ethers – a case study

The synthesis of the crown ethers and the recognition of their unique properties played a very important role in the development of macrocyclic chemistry. The compounds were discovered and developed by C.J. Pedersen. The original discovery was purely accidental and followed the isolation of 0.4% of dibenzo-18-crown-6 from the reaction of catechol and $Cl(CH_2)_2O(CH_2)_2Cl$ (Fig 3.7). Subsequent optimization of the reaction conditions lead to yields of the order of 45%. Pedersen recognized that alkali metal salts behaved strangely in the presence of this, and other, crown ethers and commenced a systematic investigation of their properties.

Pedersen was awarded the Nobel prize for Chemistry in 1987 together with D.J. Cram and J.-M. Lehn

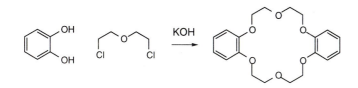

Fig. 3.7. The original synthesis of dibenzo-18-crown-6.

The basic synthetic strategy for crown ethers remains similar to the original Pedersen synthesis although it is more logical to try to use a [1+1] strategy rather than the original [2+2] condensation. The basic strategy is the nucleophile-electrophile approach presented earlier and relies upon the reaction of a diol or diphenol with a bis(electrophile). The electrophile is usually a chloro or bromo compound or a tosylate. The reactions require strong bases to be present in order that the alcohols are deprotonated and their nucleophilicity increased. The strategy applied to the synthesis of 18-crown-6 is illustrated in Fig. 3.8.

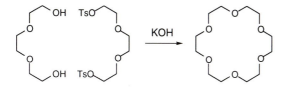

Fig. 3.8. The synthesis of 18-crown-6 from a [1+1] cyclization.

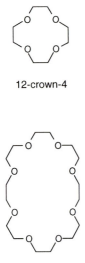

12-crown-4

24-crown-8

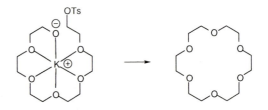

15-crown-5

These reactions do not require excessively high dilution and it is pertinent to ask why high yields of the desired crown ethers are obtained. The answer to this question leads us neatly into the next chapter.

The condensation of $Cl(CH_2)_2O(CH_2)_2Cl$ with $HO(CH_2)_2O(CH_2)_2OH$ in the presence of base could give the [1+1] product 12-crown-4, the [2+2] product 24-crown-8 or higher cyclic or acyclic [n+n] oligomers. The yields of the two crown ethers were found to be dependent upon the base that was used. With LiH as base, considerably more of the smaller 12-crown-4 was obtained whilst with NaOMe, the major product was 24-crown-12. Hole-size arguments suggest that 12-crown-4 is a better ligand for lithium than for sodium and raises the possibility of the coordination of the metal ion to the product or the intermediates controlling the reaction.

Similar effects are observed in the synthesis of 15-crown-5 from $Cl(CH_2)_2O(CH_2)_2Cl$ with $HO(CH_2)_2O(CH_2)_2O(CH_2)_2OH$ in the presence of base. A number of significant observations were made. Firstly, the yield of 15-crown-5 was at a maximum when NaOH was used as the base. Indeed, yields of the order of 40% were reported. We established in chapter 2 that sodium cations were the best match for the cavity in 15-crown-5. The use of LiOH or KOH resulted in much reduced yields of the desired crown ether (4% and 20% respectively). Even more significantly, the use of a tetraalkylammonium hydroxide as base, which has a similar base strength to the metal hydroxides, gave virtually no crown ether. Once again, the role of the metal ion needs to be clarified.

Why should these cyclizations proceed at high concentrations? It is suggested that the metal cation, which is capable of binding to the hard oxygen donor atom, pre-organizes the intermediate so that cyclization is favoured. This is illustrated in Fig. 3.9 for the formation of 18-crown-6 with a potassium salt as base. With this observation we end this chapter and consider its further implications in Chapter 4.

OTs

Fig. 3.9. The role of the metal ion in pre-organizing the intermediate prior to cyclization in the synthesis of 18-crown-6.

4 Synthetic aspects 2. The complexes

4.1 Non-template synthesis of complexes

The simplest method of preparing a metal complex is to treat the ligand of choice with a derivative of the metal of choice in an appropriate solvent. This is the traditional method of coordination chemistry and is highly effective when the rate of complexation of the ligand is reasonably fast and when the ligand itself is stable under the reaction conditions.

In practice, this approach is used for the synthesis of the majority of complexes of crown ethers, aza and thia macrocycles, many mixed donor ligands and ligands such as porphyrin and phthalocyanine.

One point of some importance is the rate of coordination of the ligand to the metal, which may be quite slow, and long reaction times are necessary before equilibrium is reached.

The nature of any anions which are present is also of some importance and we saw in Chapter 2 that crown ether complexes with alkali metal ions exhibited structures which varied if anions were coordinated to the metal centres. Similarly, the reaction of nickel(II) chloride with cyclam in aqueous solution gives an equilibrium mixture containing yellow square-planar $[Ni(cyclam)]^{2+}$ and blue $[Ni(cyclam)Cl_2]$ species. The ratio of the square-planar to octahedral complexes depends upon the anion and also upon the temperature and solvent. Recrystallization of the reaction mixture may selectively enrich one of the compounds in the solid state.

However, the direct combination method allows the preparation of some unique and interesting compounds. The crown ethers possess hard oxygen donors, and although we normally think of their coordination chemistry in terms of Groups 1 and 2, they also form complexes with lanthanoids and (less commonly) with d-block metal ions. A typical example will illustrate some of the features of direct combination reactions. Europium(III) compounds are of interest as they are luminescent and many applications are being investigated in which europium complexes are components of photoresponsive molecular machines. The hydrogenation of dibenzo-18-crown-6 gives the saturated dicyclohexano-18-crown-6 ligand **4.1**. The reaction of **4.1** with $Eu(NO_3)_3$ gives a luminescent compound containing the $[Eu(\mathbf{4.1})(NO_3)_2]^+$ cation (Fig. 4.1). However, the counter-ion is not the nitrate that might be expected but the $[Eu(NO_3)_5]^{2-}$ anion and the complex has the overall stoichiometry $([Eu(\mathbf{4.1})(NO_3)_2])_2[Eu(NO_3)_5]$. Numerous other crown ether complexes with lanthanoids are known and are invariably prepared by direct reaction of the metal salt with the desired ligand.

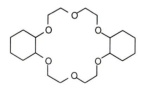

dicyclohexano-18-crown-6

(4.1)

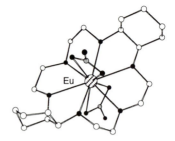

Fig. 4.1. The structure of the $[Eu(\mathbf{4.1})(NO_3)_2]^+$ cation in $([Eu(\mathbf{4.1})(NO_3)_2])_2[Eu(NO_3)_5]$.

This latter compound raises a general point regarding the relationship between stoichiometry and structure. The fact that a compound contains a metal centre and a macrocyclic ligand does not necessarily mean that the metal is to be found within the macrocyclic cavity. A good example is found in the compound $(\mathbf{4.1})_3(H_2O)_6(OH)_2(O)_2U_2(ClO_4)_2$ obtained from **4.1** and uranyl salts. The compound contains the diuranium(VI) cation $[(H_2O)_3(O)_2U(\mu\text{-}OH)_2U(O)_2(H_2O)_3]^{2+}$ and free **4.1** molecules in the solid state.

Even if the metal is coordinated to the macrocycle, it is not necessarily within the cavity. We have seen some examples in Chapter 2 where a metal ion lies above the macrocyclic cavity, but sometimes more extreme cases are found. It is constructive to consider the behaviour of $[14]aneS_2$ with mercury(II) compounds. The 1:1 adduct with mercury(II) perchlorate contains the cation $[Hg([14]aneS_4)(H_2O)]^{2+}$ in which the metal ion is more or less within the macrocyclic cavity. A 1:1 adduct is also formed with mercury(II) iodide, but in this case a polymer is formed in which HgI_2 units are bonded to one sulfur donor of each of two macrocyclic ligands. Each $[14]aneS_4$ ligand has two exocyclic bonds to mercury (Fig. 4.2 a). It is also possible to isolate 2:1 adducts such as $[Hg_2([14]aneS_4)Cl_4]$ in which each mercury centre is four coordinate and bonded in an exocyclic manner to two sulfur atoms from the macrocycle (Fig. 4.2 b).

$[14]aneS_4$

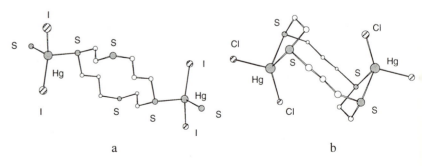

a   b

Fig. 4.2. The structures of (a) the polymer $[\{HgI_2([14]aneS_4)\}_n]$ and (b) $[Hg_2([14]aneS_4)Cl_2]$.

There are a number of rather unusual applications of direct coordination reactions of crown ethers and cryptands. Although it is usual to think of alkali metals having a cationic chemistry because of the low first ionization energy (492 kJ mol^{-1} for sodium), these elements also have favourable electron attachment energies (−53 kJ mol^{-1} for sodium). This value of −53 kJ mol^{-1} might seem to be rather modest, but should be compared with a value of −73 kJ mol^{-1} for the formation of the hydride ion. This suggests that salts containing anions such as Na$^-$ might be isolable if suitable counterions are available. This is indeed the case, as illustrated in Fig. 4.3. The treatment of solutions of Group 1 metals in ammonia or other inert solvents with crown ethers or cryptands results, in suitable cases, in the formation of alkalide salts. The alkalide salts are very air- and water- sensitive but are otherwise relatively stable. The compounds themselves have beautiful metallic appearances.

Na$^-$ is the sodide ion.

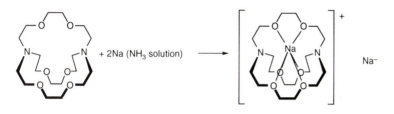

Fig. 4.3. The formation of a sodide salt by the reaction of a sodium solution with [2.2.2].

The constitution of solutions of Group 1 metals in ammonia and related solvents is very concentration dependent, but under some conditions they are best regarded as containing solvated M$^+$ cations and solvated electrons. Is it possible to isolate compounds in which the anion is simply an electron? Once again the answer is yes, and the reaction of caesium with 18-crown-6 gives a compound of stoichiometry {Cs(18-crown-6)$_2$}. The X-ray structure reveals a sandwich [Cs(18-crown-6)$_2$] unit with normal Cs-O distances for a Cs$^+$ cation. The compound is an *electride* in which the counter ion is an electron and correct formulation is [Cs(18-crown-6)$_2$]$^+$e$^-$.

A related use of cryptands is in the formation of discrete salts containing Zintl ions. Zintl phases are typically formed from the reaction of Group 1 elements with Group 14 or 15 elements and contain main group cluster anions. A typical example is the compound Na$_3$Sb$_{11}$ which is a metallic looking solid. Treatment with the cryptand [2.2.2] allows the isolation of the discrete salt [Na([2.2.2])]$_3$[Sb$_{11}$] (Fig. 4.4).

Fig. 4.4. The [Sb$_{11}$]$^{3-}$ anion present in [Na([2.2.2])]$_3$[Sb$_{11}$].

Many of the applications of macrocyclic ligands are based upon direct coordination interactions. Examples include the design of ligands for the selective extraction of removal of specific metal ions and the development of sensors for specific metal ions. In the case of extraction, the usual approach is to treat an aqueous solution containing the metal ion of interest mixed with other metal ions with a solution of the ligand in a non-aqueous phase. Hopefully, selective binding of the metal ion to the ligand will occur and the metal of interest will be selectively extracted into the non-aqueous phase.

Subsequent back-extraction into the aqueous phase, usually by a change in the pH or halide ion concentration of the aqueous phase, allows the metal to be isolated. The detection of Group 1 ions can be achieved if a crown ether bears some other functional group which registers the change on passing from the free ligand to the complex. A typical example of a ligand designed as such a sensor is seen in **4.2**; in this ligand the ferrocene shows a well-resolved and reversible Fe(II)/Fe(III) redox process. This redox process is expected to change upon the binding of a cation to the crown ether.

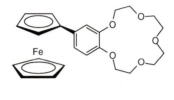

(**4.2**)

4.2 Template synthesis – metal-directed reactions

The template effect was discovered in the 1960's and has been widely used for the synthesis of macrocyclic complexes ever since.

We saw in Section 3.1 that one of the problems with the synthesis of macrocycles was the conformation control in bringing the two ends of a chain together in the final cyclization step. The *template effect* is a widely used strategy which circumvents these problems. In essence, the method involves the incorporation of additional donor atoms into the chain and performing the cyclization reaction in the presence of a metal ion which can coordinate to these. The idea is that the metal ion will coordinate to the donor atoms and pre-organize the various intermediates in the conformation required to give the desired cyclic products. This is shown schematically in Fig. 4.5 and a specific example was introduced in Fig. 3.9.

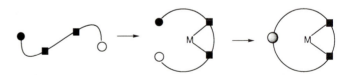

Fig. 4.5. A schematic view of the cyclization step involved in a templated macrocycle synthesis. The filled and open circles represent mutually reactive functional groups and the squares are additional donor atoms. The binding of the metal to the donor atoms pre-organizes the ligand into the conformation required for cyclization.

Template reactions may involve any number of reactants, although [1+1] or [2+2] cyclization reactions are probably the most common. It is very often found that the metal ion remains coordinated to the macrocycle and that a metal complex is obtained directly (although this was not the case in the reaction presented in Fig. 3.9).

One of the first deliberate template reactions is presented in Fig. 4.6, and this serves to illustrate a number of the important features. Firstly, the metal ion remains coordinated and a nickel(II) complex of the N_2S_2 macrocycle is obtained. Secondly, the ring forming reactions are usually similar to those used in metal free syntheses – in this case the electrophile-nucleophile strategy is adopted. Thirdly, and in this case most importantly, it is not possible to perform this reaction in the absence of a metal ion as the required reactant **4.3** cannot be isolated. Attempts to obtain **4.3** from the reaction of biacetyl (MeCOCOMe) with $H_2NCH_2CH_2SH$ result in the formation of polymers and the cyclic species **4.4**. However, if the biacetyl is reacted with the nickel(II) complex of $H_2NCH_2CH_2SH$, or nickel(II) salts are reacted with the cyclization product **4.4**, the nickel(II) complex of **4.3** is obtained directly. This emphasizes a very important point – it is possible to use template reactions when the organic precursors themselves are not stable or when the metal-free macrocyclic ligand cannot be prepared.

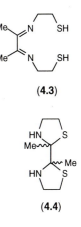

(4.3)

(4.4)

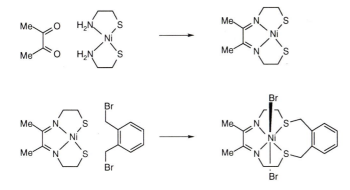

Fig. 4.6. The template synthesis of a nickel(II) macrocyclic complex.

The metal ion plays a number of roles in the above reaction. The stabilization of the acyclic intermediate **4.3** is achieved in a number of ways. As a free ligand, **4.3** possesses nucleophilic thiol and electrophilic imine groups and cyclization arises by attack of the thiol upon the carbon of the imine to give **4.4**. When **4.3** is coordinated to the nickel(II), the nucleophilicity of the thiolate is reduced because it is bonded to an electropositive centre, the thiolate is constrained so that it cannot approach the imine, and the imine is made less electrophilic by the back-donation of *d*-electron density into the π^* orbitals. The metal also pre-organizes the two thiolate groups so that they are correctly oriented for reaction with the electrophilic 1,2-bis(bromomethyl)benzene to give the macrocyclic ligand. One of the more subtle effects of the metal ion is to reduce the *reactivity*, and hence increase the *selectivity*, of the thiolate groups. Thus, although the cyclization reaction is substantially slower than the reactions leading to **4.4**, the selectivity for the macrocyclic product is very high.

The advantages of template reactions are obvious (good yields, obtaining metal complexes directly, mild reaction conditions), but it must be stressed

Imines are electrophilic. The empty π^* orbitals have a greater coefficient at carbon and this is the favoured site of attack. Upon coordination to an electron-rich transition metal ion, back-donation of electrons results in an increasing electron density within these orbitals and nucleophilic attack is disfavoured. The result is that imines are very often stabilized upon coordination to a metal ion.

that there are also disadvantages. The major disadvantage is that not all metal ions will act as a template for a desired reaction and it is often a rather imprecise art in finding the correct metal ion. This creates rather obvious problems if you are interested in the iron complexes of a macrocyclic ligand but only nickel functions as a template. Although metal exchange reactions are sometimes possible, these are usually controlled by the Irving-Williams order of stability in which nickel(II) complexes are close to the maximum stablization. The conditions required to remove metal ions from macrocyclic complexes can often lead to fragmentation of fragile ligand systems and it is not always possible to convert templated products to the free ligands.

Another disadvantage is that it is not always possible to predict exactly what the product might be. In the reaction in Fig. 4.6, only the [1+1] macrocyclic product is likely on steric grounds, but we saw in Section 3.4 that the reaction of $Cl(CH_2)_2O(CH_2)_2Cl$ with $HO(CH_2)_2O(CH_2)_2OH$ gave varying amounts of 12-crown-4 and 24-crown-8 depending upon the base used. Although we can now recognize this as a template reaction, it seems that the coding inherent in the matching of the metal ion size to the macrocyclic cavity is relatively weak. A particularly impressive example of template effects in the synthesis of crown ethers is seen in the oligomerization of oxirane (ethylene oxide) by BF_3 in the presence of various Lewis acidic metal salts. In the absence of the metal salts, the yields of cyclic oligomers are low, with 1,4-dioxan dominating the products. However, when the metal salts are present, good yields of crown ethers are obtained and there is a good relationship between the hole-size and the metal ionic radius (Fig 4.7).

1,4-dioxan

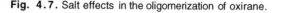

Salt	%	%	%
none	15	5	4
LiBF$_4$	30	70	-
NaBF$_4$	25	50	25
KBF$_4$	-	50	50
RbBF$_4$	-	-	100
CsBF$_4$	-	-	100

Fig. 4.7. Salt effects in the oligomerization of oxirane.

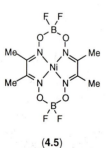

(4.5)

With template effects involving Group 1 metal ions, it is usually the free macrocycle that is isolated, but with transition metal templated reactions it is more often found that stable metal complexes are formed. A typical example is **4.5** which is the product of the reaction of **4.6** with $BF_3 \cdot OEt_2$. The macrocycle in **4.5** is dianionic with two formally negatively-charged boron centres.

One of the most common templated reactions involves the formation of imine bonds. Pyridine aldehydes and ketones are widely-used carbonyl compounds in such reactions as they incorporate additional donor capacity within the macrocyclic ring and a typical example of a template reaction of this type is presented in Fig. 4.8. This ligand represents one of a very large series obtained from the template condensation of 2,6-diacetylpyridine or 2,6-pyridinedicarboxaldehyde with amines of the general formula $H_2N(CH_2)_m\{NR(CH_2)_n\}_pNH_2$. Subsequent studies involved the incorporation of pyrrole, thiophene or furan rings instead of the pyridine and the use of α,ω-diamines bearing pendant functionality. The macrocyclic ligands **2.4** and **2.7** are also obtained in reactions utilizing nickel(II) templates.

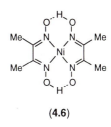

(4.6)

α,ω-Diamines have two terminal amino groups linked by a chain.

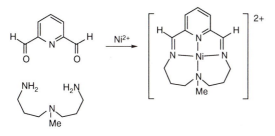

Fig. 4.8. A typical nickel(II) template reaction in which two imine bonds are formed.

Although tetraaza macrocycles are usually prepared in metal-free syntheses, the simplest preparation of $[Ni(cyclam)]^{2+}$ is by the template reaction in Fig. 4.9. The reduction of the imine bond is readily achieved and subsequent treatment with an excess of potassium cyanide results in the removal of nickel as $Ni(CN)_4]^{2-}$ and the cyclam may be extracted. A similar reduction and demetallation of the nickel(II) complex of the Curtis macrocycle **3.12** leads to the saturated compound **2.42**.

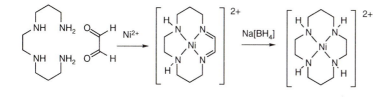

Fig. 4.9. A template reaction used in the synthesis of $[Ni(cyclam)]^{2+}$.

We have already seen that **3.12** may be prepared in a metal-free reaction, but the nickel(II) complex is more conveniently prepared in a template reaction. All that is necessary is to heat purple $[Ni(en)_3][ClO_4]_2$ in acetone – the solution turns yellow and the complex $[Ni(\mathbf{4.7})][ClO_4]_2$ may be isolated. It is interesting to note that in this metal-templated process, both $[Ni(\mathbf{4.7})]^{2+}$ and isomeric $[Ni(\mathbf{4.8})]^{2+}$ salts are isolated. This remarkable

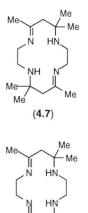

(4.7)

(4.8)

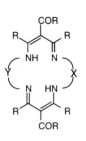

The generic structure of a Jäger macrocycle.

template reaction involves four molecules of acetone and two of 1,2-diaminoethane and proceeds through the formation of C–N, C=N and C–C bonds!

It is possible to form more than two imine bonds in a template condensation, and typical examples are seen in **4.9** which is obtained from the nickel(II)-templated condensation of 1,3-propanediamine hydrochloride with biacetyl and **4.10** which is prepared in another nickel(II) templated process, this time involving pentane-2,4-dione and 1,2-diaminobenzene. Both of these macrocyclic ligands are examples of [2+2] condensation products and the ligand in **4.10** is doubly deprotonated to give an overall neutral nickel complex. The ligand in **4.10** is one of a large class of compounds which are generically known as Jäger macrocycles – these very often incorporate additional carbonyl groups.

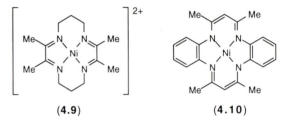

(4.9)   (4.10)

Template reactions are not limited to compounds with nitrogen, oxygen or sulfur donors and recent extensions of the method have allowed the synthesis of phosphorus donor macrocyclic complexes (Fig. 4.10). Note that in this case, the dehydration to a multiply-bonded analogue of an imine does not occur.

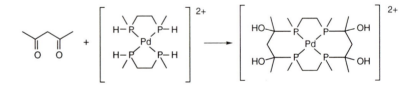

Fig. 4.10. The template synthesis of a palladium complex with a P_4 macrocyclic ligand.

4.3 Types of template effect

Two different types of template effect have been described. The first is the *kinetic template effect*. In this, the metal plays essentially the role which we have assigned it above. The function of the metal is to control the stereochemistry in the intermediates such that cyclization is the favoured pathway. Implicit in the kinetic template effect is the concept that the macrocycle of interest is not formed from the same reactants in the absence of a template. The reaction presented in Fig. 4.6 is an example of the kinetic template effect.

The second type of template effect which has been described is known as the *thermodynamic template effect*. In a reaction of this type, the macrocyclic product is formed, along with other species, in the absence of the templating metal ion. The function of the metal ion is to coordinate to the macrocycle and to remove it from the equilibrium mixture. The formation of the nickel(II) complex of the Curtis macrocycle is an example of the thermodynamic template effect, as we saw in Chapter 3 that the free macrocycle could be obtained in a metal-free reaction.

Whilst these two types of template effect appear very clear in principle, the majority of template reactions have not been investigated in sufficient detail to allow a precise classification. The interest in template reactions is more often preparative rather than mechanistic and detailed studies of the mechanism of such reactions are few and far between.

Two types of template effect may be described. However, in many cases the mechanistic details of the reactions are not known in detail.

4.4 Hole-size effects

We have already seen the influence of the metal ion over the product distribution in the formation of crown ethers from oxirane. In this section we look at hole-size effects involving transition metal ions.

Some of the best examples arise from the choice of [1+1], [2+2] or [n+n] products in reactions of the type presented in Fig. 4.8. The condensation of 2,6-diacetylpyridine with $H_2N(CH_2)_3NH(CH_2)_3NH_2$ in the presence of nickel(II) salts yields [Ni(**4.11**)]$^{2+}$, a complex of the [1+1] macrocycle which is approximately the correct size for the nickel(II) ion. When the same reactants are condensed in the presence of silver(I) salts, a different reaction occurs. The silver(I) ion is too large for the cavity in the [1+1] macrocycle, and the product of the reaction is a disilver complex of the [2+2] macrocycle **4.12**. Although the above rationalization appears satisfactory, it is not really clear why the [2+2] macrocycle is formed, as the silver ion is actually too small for the cavity of **4.12**, and this results in the formation of a *dinuclear* complex.

The more these systems are studied, the less predictable they become. For example, the condensation of 2,6-diacetylpyridine with $H_2NCH_2CHOHCH_2NH_2$ in the presence of manganese(II) might be expected to yield a [2+2] macrocyclic system; however, the product of the reaction is the tetramanganese complex of the [4+4] macrocyclic ligand **2.40**.

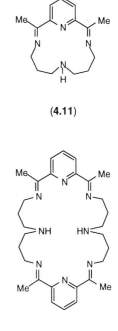

(**4.11**)

(**4.12**)

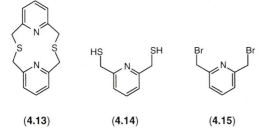

(**4.13**)   (**4.14**)   (**4.15**)

In Chapter 2 we noted that it was not possible to define an unambiguous hole-size for a macrocyclic ligand. It is worth emphasizing this point again. In choosing a template ion for a desired reaction, it seems logical to select a metal ion which is the correct size for the cavity of the new macrocycle. For example, the 12-membered macrocyclic ligand **4.13** is expected to be too small for nickel(II) on the basis of the hole-size and a different metal ion would be expected to act as a template. However, the template condensation of **4.14** and **4.15** proceeds smoothly in the presence of nickel(II) salts. The explanation lies in the ability of the flexible macrocycle to adopt a folded conformation in the complex [Ni(**4.13**)Br$_2$], in which the two bromide ligands occupy *cis* positions.

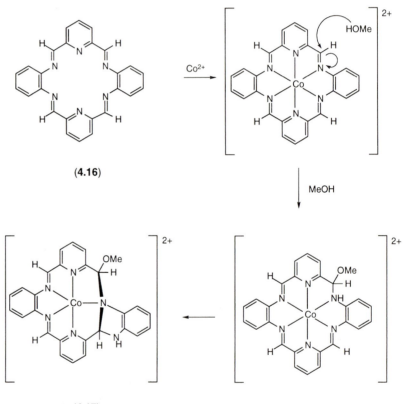

(4.16)

(4.17)

Fig. 4.11. A dramatic rearrangement that occurs when **4.16** is treated with cobalt(II) salts. The driving force is the formation of a macrocyclic cavity which can accommodate cobalt(II).

One of the more intriguing aspects of macrocyclic chemistry is the ability of the ligands to undergo substantial rearrangement to give optimal metal-binding cavities. Many examples of such rearrangements are known, but those involving the condensation of α,ω-diamines with dicarbonyl compounds are particularly well-studied. The 18-membered [2+2] macrocyclic ligand **4.16** is prepared in a metal-free condensation of 2,6-pyridinedicarboxaldehyde with 1,2-diaminobenzene. The cavity is too large for a first-row transition metal ion and reaction with cobalt(II) salts in methanol results in a ring-contraction to give **4.17**. The addition of the methanol results in a distortion of the ligand from planarity, but this is not enough to optimize the cavity for the cobalt(II) ion and a second step involving an intramolecular attack of the amine on the imine occurs to generate the smaller 15-membered ring macrocycle (Fig. 4.11).

Sometimes the rearrangements are so drastic that it is hard to recognize the original structural components in the product! For example, when 2,6-diacetylpyridine (instead of 2,6-pyridinedicarboxaldehyde) is reacted with 1,2-diaminobenzene, the product is **4.18** rather than the analogue of **4.16** with methyl groups. This is believed to be a result of steric interactions between the methyl groups and the phenyl *ortho*-protons. Treatment of **4.18** with copper(II) salts leads to **4.19** which contains a 15-membered macrocyclic ring.

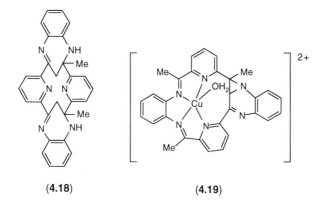

(4.18)   (4.19)

It should be apparent from this discussion that many of the reactions of imine macrocycles are, to some extent, reversible. This suggests that it should also be possible to interconvert [1+1] and [2+2] macrocycles by treatment with an appropriate metal ion. Once again, a considerable degree of subtlety is exhibited in these reactions. Compound **4.20** is a cyclic trimer of 2-aminobenzaldehyde and upon reaction with nickel(II) salts it rearranges to give nickel(II) complexes of the cyclic trimer **4.21** and the tetramer **4.22**.

The trimer **4.21** behaves like a rigid analogue of triazacyclonane and forms complexes in which it occupies the *facial* positions in an octahedron. The complex [Ni(**4.21**)(H$_2$O)$_3$]$^{2+}$ is of some interest as it is chiral (Fig. 4.12). Although the water ligands exchange with bulk water extremely rapidly, the rate of racemization is extremely slow and the resolved complex may be kept in solution for several months with negligible loss of optical activity. The tetramer behaves as a planar tetradentate ligand resembling a porphyrin or a phthalocyanine and forms square-planar or octahedral complexes in which it occupies the equatorial sites.

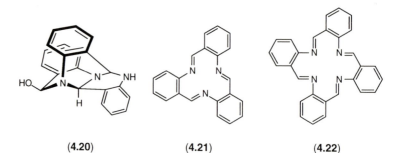

(4.20)   (4.21)   (4.22)

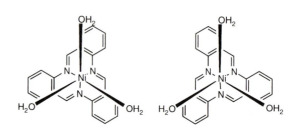

Fig. 4.12. The two enantiomers of the [Ni(**4.21**)(H$_2$O)$_3$]$^{2+}$ cation.

4.5 Assembly from more than two components

The template reactions that we have considered so far are simple in that it is possible to make a reasonable guess what the product will be, although as we have seen above, hole-size mis-matches may lead to rearrangements. In this section we will consider some less intuitive template reactions.

The prototype reaction involves the copper(II) complex **4.23** with formaldehyde and nitroethane, which yields **4.24** (Fig. 4.13). Although an exact mechanism cannot be proposed, the structure of the product may be rationalized by considering the formal involvement of the bis(imine) **4.25**. Attack of the *aci* form of nitroethane or its anion upon **4.25** can lead to **4.24**.

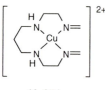

(4.25)

the *aci* form of nitroethane

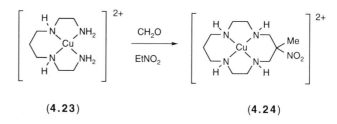

(4.23)   (4.24)

Fig. 4.13. The template synthesis of a cyclam-like macrocyclic complex.

The formation of the cyclam-like macrocycle suggests that this synthetic route may be of some interest and numerous related examples are known. It is not necessary to use nitroethane as the nucleophile and the reaction of the nickel(II) analogue of **4.23** with formaldehyde and methylamine gives **4.26**. It is even possible to start with much simpler complexes and the reaction of [Cu(en)₂]²⁺ with formaldehyde and methylamine gives **4.27**.

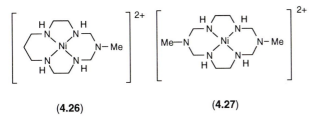

(4.26)   **(4.27)**

   These reactions are not simple and it is not always possible to make reasonable predictions of their outcome. A good example of this is the formation of **4.28** from the reaction of 1,2-diaminoethane with formaldehyde and ammonia in the presence of nickel(II) salts!

   The primary use of such reactions is the preparation of transition metal complexes of encapsulating ligands. We saw earlier that complex **4.25** was obtained from the reaction of two coordinated dimethylglyoximato ligands in **4.26** with BF_3. If the same reaction is performed with [Co(Hdmg)₃] (H_2dmg = HON=C(Me)C(Me)=N(OH)), the product is the cobalt(III) complex **4.29** in which the metal ion is trapped within a new encapsulating ligand. A related reaction is seen between the iron(II) complex [Fe(H₂NN=C(Me)C(Me)=NNH₂)₃]²⁺ (the *bis*hydrazone of biacetyl) and formaldehyde to give **4.30**. The reaction proceeds by the formation of an imine followed by nucleophilic attack by the adjacent amino group. The metal ions preorganize the ligands so that the three-dimensional encapsulating ligand is formed in good yield. A natural consequence is that the metal ion is trapped within the ligand.

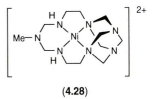

(4.28)

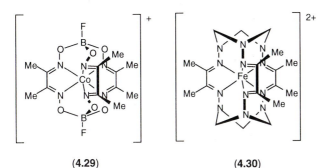

(4.29)   **(4.30)**

   Numerous examples of such condensations are known, but perhaps the most spectacular are those described by Sargeson and his co-workers. In a typical example, the complex [Co(en)₃]³⁺ is reacted with formaldehyde and ammonia to give **4.31** (Fig. 4.14). The product **4.31** is formally the reaction product of a cobalt(III) ion, three molecules of 1,2-diaminoethane, two molecules of ammonia and six molecules of formaldehyde. The

Δ and Λ are used to describe the absolute configuration of inorganic compounds and correspond to left- or right-handed orientations of tris(chelate) complexes.

An aminol is the hydrated form of an imine, $R_2C(OH)NH_2$.

encapsulating ligand that is formed is known as a *sepulchrate*, and the yields are remarkably good. This reaction is remarkable because a total of six new stereocentres are generated (one at each of the nitrogen atoms of the 1,2-diaminoethane ligands). If enantiomerically pure *Δ*- or *Λ*- $[Co(en)_3]^{3+}$ is used in these reactions, a single diastereomer of **4.31** is obtained. The mechanistic origins of this remarkable stereospecificity have been elucidated and, although they will not be presented in detail here, follow from a sequence of metal controlled reactions involving the formation and subsequent reaction of imines and aminols.

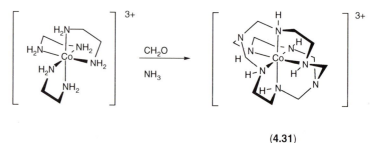

(**4.31**)

Fig. 4.14. The condensation of $[Co(en)_3]^{3+}$ with formaldehyde and ammonia gives the cobalt(III) complex of the sepulchrate ligand.

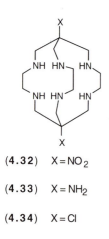

(**4.32**) X = NO$_2$

(**4.33**) X = NH$_2$

(**4.34**) X = Cl

Similar reactions occur when the ammonia is replaced by nitromethane in the above reaction and the cobalt(III) complex of **4.32** is obtained. These ligands are described as *sarcophagines* and the specific example in **4.32** is dinitrosarcophagine. These complexes have a rich chemistry and a wide variety of compounds may be prepared by standard transformations of the substituents. For example, **4.32** may be reduced to a diamino compound **4.33**, which may, in turn, be diazotized and converted to the dichlorocompound **4.34**. Although these reactions are of great synthetic value, they do not occur with all octahedral metal centres and it is extremely difficult to remove the metal from the complex to give the free ligand.

4.6 Catenanes

A final example of a template effect illustrates the elegance of the method and the high degree of control which may be exerted. A *catenane* is a molecule in which two or more rings are inter-linked, in a manner resembling a daisy chain. The problem in the synthesis of such molecules by conventional methods comes back to the conformational control that we discussed in the previous chapter. In order to form a catenane rather than two separate, non-linked, rings it will be necessary to have an intermediate state in which one of the rings has formed and the second open-chain precursor is threaded through this ring. This arrangement is called a *rotaxane*.

Extremely elegant approaches have been developed for the synthesis of catenanes, but the most successful involve the use of copper(I) templates. A tetrahedral metal complex containing two didentate ligands is the key to the synthesis. If the two didentate ligands have reactive functionality which can

Catenate

Catenand

react with another reagent to give a macrocycle it should be possible to chose the correct didentate ligand and the correct reagent such that two ends of the same ligand are linked together. This leads to the metal-coordinated catenane, called a *catenate*. This is the approach which has been developed by Sauvage and co-workers and is illustrated in Fig. 4.15.

When the correct combination of reagents is taken, the yields of the catenane can be very high. It is even possible to remove the metal ion from the catenate by reaction with cyanide to leave the inter-linked metal-free catenane (or *catenand*).

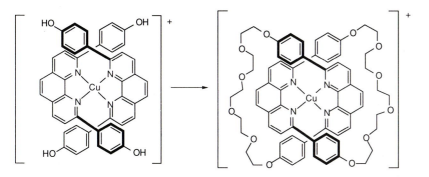

Fig. 4.15. The synthesis of a catenate. The metal arranges the bis(phenols) so that reaction with $ICH_2(CH_2OCH_2)_4CH_2I$ results in the formation of the two inter-linked rings.

The product in Fig. 4.15 is called a *2-catenane* because it contains two inter-linked rings. Developments of this strategy have allowed the synthesis of 3-catenanes and higher analogues. Perhaps the most elegant extension of this synthetic method is seen in the synthesis of a molecular knot about transition metal templates. The product **4.35** in Fig 4.16 is a *trefoil knot* and the synthesis follows from the correct orientation of the ligand about two copper centres so that reaction with the bis(electrophile) leads to the knotted rather than any of the possible catenanted or macrocyclic products. Further discussion of these reactions is beyond the scope of this book.

A trefoil knot.

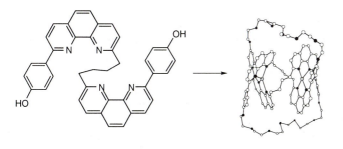

(**4.35**)

Fig. 4.16. The formation of a trefoil knotted complex from the reaction of an open chain dinucleating ligand with copper(I) followed by $ICH_2(CH_2OCH_2)_5CH_2I$.

5 Thermodynamic and kinetic considerations

5.1 The macrocyclic effect

The chelate effect states that complexes with multidentate ligands are more stable than those with an equivalent number of similar monodentate ligands. This apparently simple observation is complicated when it is realized that the dimensions of K differ between chelating and monodentate ligands.

Complexes with multidentate ligands are more stable than those containing the same number of equivalent monodentate ligands. It is an obvious extension to ask if cyclic ligands give more stable complexes than corresponding open-chain ligands. This is indeed the case, and the phenomenon is described as *the macrocyclic effect*. Many of the applications of macrocyclic ligands rely upon this macrocyclic effect and in this chapter we shall examine the origins and consequences of this additional stability.

5.2 Quantification of the macrocyclic effect

The macrocyclic effect states that complexes with macrocyclic ligands are more stable than those with equivalent open-chain ligands.

The macrocyclic effect was first quantified for the copper(II) complex of ligand **5.1** (Eqn. 5.1) which was found to be about 10^4 times more stable than the corresponding complex with the open chain tetraamine **5.2**. It was

$$\mathbf{5.1} + Cu^{2+} \rightarrow [Cu(\mathbf{5.1})]^{2+} \qquad \lg K = 28 \qquad (5.1)$$

also noted in these early studies that the rate of coordination of **5.1** was very much slower (10^3-10^4 times) than that of **5.2**. This observation has proved to be general, and macrocyclic compounds are almost invariably more stable than those with equivalent open-chain ligands. However, with very large, flexible rings containing large numbers of donor atoms, the effect becomes insignificant.

Before we discuss the detailed origins of the macrocyclic effect, we should emphasize the experimental difficulties that are encountered in such studies. Reaction systems may take days or weeks to reach equilibrium, pushing calorimetric methods to their limits. With some metal ions, various solution species may exist, differing in coordination number and/or spin state. Another problem lies in the comparison with 'equivalent open-chain ligands'. Clearly the donor atoms should be the same in the macrocyclic and open-chain species – it would be pointless to compare oxygen donors with phosphorus donors. However, a number of more subtle points also need to be addressed regarding the nature of the spacer groups between the donor atoms. While **5.2** is probably a good model for **5.1** or cyclam, and **5.3** will represent cyclen, would **5.4**, **5.5** or **5.6** be better for [15]aneN$_4$? If we represent the pattern of methylene groups within [15]aneN$_4$ as 2333, then **5.4** presents a 233 set, **5.5** a 323 set and **5.6** a 333 set.

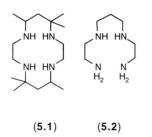

(5.1)   (5.2)

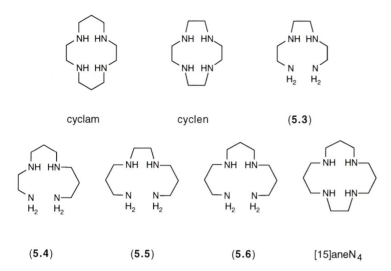

cyclam   cyclen   (5.3)

(5.4)   (5.5)   (5.6)   [15]aneN$_4$

(5.7)

Furthermore, is it reasonable to compare cyclam, with four secondary amines, to **5.1**, with two secondary and two primary amino groups, or would **5.7** be a better model? In the detailed discussion that follows, tetraazamacrocycles will be discussed, but the general arguments apply equally to other donor systems.

5.3 The origin of the macrocyclic effect

It is very tempting to suggest that the macrocyclic and chelate effects have a common origin in the entropic term. Unfortunately, this has not proved to be the case, and, after a long period of acrimonious discussion, it is now clear that it is not possible to ascribe a single origin to the macrocyclic effect. Indeed, the best summary comes from Paoletti who states "the macrocyclic effect is an experimental observation but its correct quantitative evaluation is an arduous task."

Much of the early confusion arose because thermodynamic determinations were not made under similar solvent or ionic strength conditions; indeed, it is now known that solvation effects are critical and that, for example, the differences between 9:1 and 9.5:0.5 MeOH:H$_2$O have critical effects upon the thermodynamic parameters. In addition, early studies of nickel(II) complexes with cyclam and related ligands did not fully appreciate that aqueous solutions contained both low-spin, yellow, square planar [NiL]$^{2+}$ and high-spin, blue, octahedral [NiL(H$_2$O)$_2$]$^{2+}$ species. Temperature-dependent studies allow ΔH and ΔS values to be obtained and Table 5.1 presents data for the high- and low-spin complexes formed by nickel(II) with cyclam and ligand **5.2** at 298 K.

$$\Delta G = \Delta H - T\Delta S$$

$$\Delta G = -RT \ln K$$

$$\frac{\mathrm{d}(\ln K)}{\mathrm{d}T} = -\frac{\Delta H}{RT^2}$$

Table 5.1. Thermodynamic data for the formation of nickel(II) complexes at 298 K.

Ligand	Low-spin		High-spin	
	ΔH / kJ mol^{-1}	$T\Delta S$ / kJ mol^{-1}	ΔH / kJ mol^{-1}	$T\Delta S$ / kJ mol^{-1}
cyclam	−78.2	49.3	−100.8	24.3
5.2	−66.1	21.7	−80.3	10.9

The data show that *both the entropic and the enthalpic effects contribute to the macrocyclic effect* (i.e. that ΔG for the formation of the cyclic complex is more negative than for the complex with **5.2**).

Table 5.2. Thermodynamic data (298 K) for $[\mathrm{CuL}]^{2+}$ complexes with N_4-donor ligands.

Ligand	lg K	ΔH / kJ mol^{-1}	$T\Delta S$ / kJ mol^{-1}
cyclen	24.8	−95.0	46.5
[13]aneN$_4$	29.1	−166.0	107.1
cyclam	27.2	−155.2	135.6
[15]aneN$_4$	24.4	−139.2	110.9
[16]aneN$_4$	20.9	−119.3	83.7
5.3	20.2	−115.3	90.4
5.2	23.9	−136.4	115.9
5.5	21.8	−124.5	108.4
5.6	17.3	−98.8	81.6
5.8	20.9	−119.3	88.3
5.7	21.89	−124.9	108.4
5.9	18.50	−105.6	93.1
5.10	14.62	−83.4	62.3

[13]aneN$_4$

[16]aneN$_4$

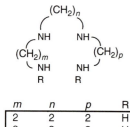

	m	n	p	R
5.3	2	2	2	H
5.2	2	3	2	H
5.5	3	2	3	H
5.6	3	3	3	H
5.8	2	2	2	Me
5.7	2	3	2	Me
5.9	3	2	3	Me
5.10	3	3	3	Me

The role of the model compound, and of ring size, may also be probed. Table 5.2 presents data for copper(II) complexes with a series of tetraaza ligands and model compounds.

The first effect to note is that when the macrocyclic ligands are compared with the open-chain analogues, both the entropic and enthalpic effects favour the macrocyclic complex. Secondly, a hole-size effect is observed with a maximum stability at the 13-membered ring in [13]aneN$_4$. Finally, the alkylated ligands behave slightly differently to the open-chain species with terminal NH$_2$ groups. As the plots of lg K show (Fig. 5.1), the alkylated ligands provide a better model for the macrocyclic species and the trends parallel each other, whereas that for the ligands with NH$_2$ groups tends to converge with the macrocycles.

Similar trends occur with crown ethers, sulfur or mixed donor-atom macrocycles upon comparison with open-chain analogues, although the relative importance of enthalpy and entropy contributions is very variable. Two typical examples are given on page 57.

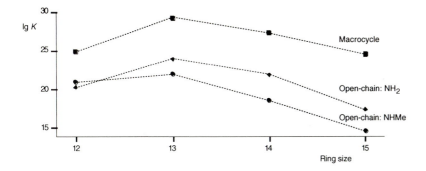

Fig. 5.1. Trends in stability constants for $[CuL]^{2+}$ complexes as a function of ring size and type of ligand.

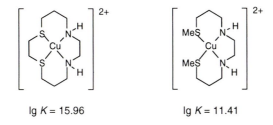

lg K = 15.96   lg K = 11.41

The series of ligands **5.11**-**5.13** all coordinate alkali metal ions, and the 1:1 complexes with potassium ions have values of lg K = 2.05, 2.27 and 6.05 respectively. Once again, the macrocyclic effect is operative and the complex with ligand **5.13** is stabilized with respect to the equivalent open-chain compounds **5.11** and **5.12**. In the case of **5.12** and **5.13** binding potassium ions, the dominant factor appears to be the enthalpy term (Table 5.3), but a comparison of the thermodynamic data for the formation of the sodium complexes of **5.12** and **5.13** indicates that it is the entropy term that is dominant! The precise balance between the various terms is dependent upon the solvent, the particular ligand under investigation and the availability of anions to coordinate to the metal centres.

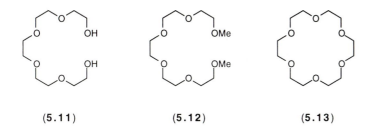

(**5.11**)   (**5.12**)   (**5.13**)

Table 5.3. Thermodynamic data for sodium and potassium complexes with ligands **5.12** and **5.13**.

	5.12			5.13		
	lg K	ΔH / kJ mol^{-1}	$T\Delta S$ / kJ mol^{-1}	lg K	ΔH / kJ mol^{-1}	$T\Delta S$ / kJ mol^{-1}
Na$^+$	1.0	−38.26	−32.23	4.33	−33.95	−9.21
K$^+$	2.27	−34.16	−21.18	6.05	−55.29	−20.76

It is rather more difficult to obtain a qualitative understanding of the *origin* of the macrocyclic effect. However, the favourable enthalpy term has been shown to be almost entirely due to the differential solvation of macrocyclic and open-chain ligands. The macrocyclic effect can be expressed in the equilibria 5.2 and 5.3, where L^1 is macrocyclic and L^2 is open-chain. If the macrocyclic complex is more stable, then the reaction should lie to the right hand side, as we know is the case for Eqn. 5.3. This approach emphasizes that the effect arises from the difference in stability of complexes with macrocyclic and open-chain ligands. Equation 5.2 relates to the exchange of macrocyclic for open-chain ligand in aqueous solution, whereas Eqn. 5.3 refers to the (hypothetical) gas phase process. If solvation is the dominant feature, then the favourable ΔH term for Eqn. 5.2 should disappear in the gas phase and Eqn. 5.3 should have an enthalpy term close to zero.

$$[ML^2]^{2+}(aq) + L^1(aq) = [ML^1]^{2+}(aq) + L^2(aq) \qquad (5.2)$$

$$[ML^2]^{2+}(g) + L^1(g) = [ML^1]^{2+}(g) + L^2(g) \qquad (5.3)$$

The enthalpy changes for Eqn. 5.2 and 5.3 are related as shown by Eqn. 5.4.

$$\Delta H(aq) = \Delta H(g) + \{\Delta_{hyd}H([ML^1]^{2+}) - \Delta_{hyd}H([ML^2]^{2+})\}$$

$$+ \{\Delta_{hyd}H(L^2) - \Delta_{hyd}H(L^1)\} \qquad (5.4)$$

Of the various terms in Eqn. 5.4, $\Delta H(aq)$, $\Delta_{hyd}H)(L^2)$ and $\Delta_{hyd}H)(L^1)$ may be experimentally determined. If the hydration energies of the two complexes are assumed to be approximately similar, the macrocyclic effect almost disappears in the gas phase. For example, comparing copper(II) complexes with cyclam and ligand **5.2**, $\Delta H(aq) = -19.7$ kJ mol^{-1}, $\{\Delta_{hyd}H(\text{cyclam}) - \Delta_{hyd}H(\textbf{5.2})\} = -19.2$ kJ mol^{-1} and $\Delta H(g) \approx 0$ kJ mol^{-1}.

A physical interpretation is that the open-chain ligand is more solvated (with a negative $\Delta_{hyd}H$) than the macrocyclic ligand. Accordingly, Eqn 5.2 but not Eqn 5.3 lies to the right-hand side. In Eqn 5.3, the metal-ligand bonds and the gas phase stabilities of the ligands are similar without regard to whether the ligand is open-chain or macrocyclic.

It is harder to give a physical interpretation to the entropic component. Cyclic ligands have less rotational and translational freedom than open-chain ligands, and so there will be a less dramatic ordering effect upon coordination of the macrocycle to a metal than the open-chain ligands. In other words, the donor atoms in macrocycle are *pre-organized* for coordination. Less easy to predict, but important, are entropy changes associated with solvation.

5.4 The cryptate effect

In the same way that macrocyclic ligands give more stable complexes than open-chain ligands, polycyclic ligands form complexes that are more stable than the corresponding macrocycles. For example, the potassium complex of 18-crown-6 is about 10^4 times more stable than that with an open-chain analogue. A similar difference in stability exists between the complexes with the cryptand [2.2.2] and a macrocyclic analogue. The precise origins of the cryptate effect have not been studied in the same detail as the macrocyclic effect, but it seems likely that a balance of entropic and enthalpic terms, together with differential solvation, is important.

18-crown-6

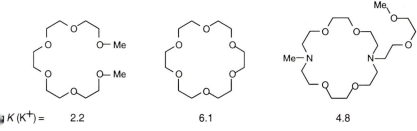

| $K (K^+) =$ | 2.2 | 6.1 | 4.8 | 9.75 |

cryptand [2.2.2]

5.5 Kinetic aspects of macrocyclic coordination chemistry

In the preceding sections, we examined the thermodynamics of coordination in some detail. We now consider the *rates* of the various reactions — are reactions with macrocyclic ligands significantly faster or slower than those with open-chain ligands? The simple answer is that both coordination and ligand dissociation reactions with macrocyclic ligands are usually slower than those with open-chain analogues. The origins and consequences of this kinetic behaviour will now be examined.

It is relevant to note that the equilibrium constant, K, for the formation of the complex is related to the rates of the forward and back reactions (Eqns. 5.5 and 5.6).

$$M^{n+} + L \; \underset{k_b}{\overset{k_f}{\rightleftharpoons}} \; ML^{n+} \tag{5.5}$$

$$K = \frac{[ML^{n+}]}{[M^{n+}][L]} = \frac{k_f}{k_b} \tag{5.6}$$

Although K for the macrocyclic ligands is considerably larger than for open-chain ligands, we cannot *a priori* say anything about the absolute values of k_f and k_b since it is only the ratio that controls stability. In practice, the rate of

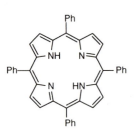

meso-tetraphenylporphyrin

formation of the complexes is 10^2-10^3 times slower than with open-chain ligands.

5.6 Formation of macrocyclic complexes

Once again, we will begin with tetraaza macrocycles. In general, the rate of formation of complexes (k_f) with the macrocyclic ligands is slower than with open-chain analogues (Table 5.4). Macrocyclic ligands with a greater steric bulk exhibit slower formation kinetics (**5.1** *versus* **5.2**), but the rate of formation is often more or less independent of the ring size (cyclen *versus* cyclam). With rigid macrocycles such as *meso*-tetraphenylporphyrin, the rates are even slower, and the copper(II) complex is formed with $k_f \approx 10^{-2}$ M^{-1} s^{-1}.

Table 5.4. Rates of formation of copper(II) complexes with tetraaza macrocycles in acidic aqueous solution.

$$Cu^{2+}(aq) + L \rightarrow [CuL]^{2+}$$

L	k_f / M^{-1} s^{-1}
5.2	10^8
cyclam	10^6
5.1	$\approx 10^4$
cyclen	$\approx 10^6$

(**5.14**)

(**5.15**)

(**5.16**)

The study of these processes with the tetraaza macrocycles is complicated by pH effects. In acidic solution, the ligands are protonated and species such as L, LH^+, LH_2^{2+}, LH_3^{3+} and LH_4^{4+} need to be taken into consideration for both thermodynamic and kinetic studies. In contrast, in basic solution, the ligands are not protonated; however, metal ions are present as various hydroxy or oxo species. These problems are to some extent circumvented by studying less basic macrocycles. The copper(II) complexes of ligands **5.14**, **5.15** and **5.16** are formed with $k_f \approx 10^6$, $\approx 10^4$ and $\approx 10^3$ M^{-1} s^{-1} respectively. Once again, the macrocyclic ligands have slower coordination rates and in this case a hole-size effect is seen.

If we now come to crown ethers and cryptands, the story is less clear, with values of k_f being more or less independent of ligand ring-size or even on going a macrocycle to a cryptand. Most studies agree that the reactions are multi-step processes, but it is not clear how the rate-determining step varies between different types of ligand. Values of k_f for the complexation of potassium ion with 18-crown-6, 15-crown-5 and the cryptands [2.2.2] and [2.2.1] are 4.3×10^8, 4.3×10^8, 4.7×10^8 and 3.8×10^8 M^{-1} s^{-1} respectively.

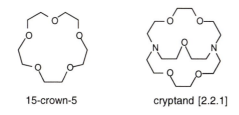

15-crown-5   cryptand [2.2.1]

How may we envisage the coordination of macrocyclic ligands to the metal ion? To summarize a great deal of painstaking and often contradictory work, the following picture emerges. The formation of a macrocyclic complex from a solvento-species MS_6 follows the *Eigen-Wilkins* or *Eigen-Winklen* mechanism. The first step involves the formation of an outer-sphere complex with a stability constant K_{OS} (Eqn. 5.7).

$$MS_6 + L \xrightleftharpoons{K_{os}} MS_6 \cdots L \qquad (5.7)$$

This is followed by an interchange process in which the first new metal-ligand bond is formed, followed by subsequent steps in which new M–L bonds are formed sequentially (Eqns. 5.8 and 5.9).

$$MS_6 \cdots L \xrightleftharpoons{k_1} MS_5L \qquad (5.8)$$

$$MS_5L \xrightarrow{k_2} \xrightarrow{k_3} \ldots \ldots \xrightarrow{k_n} ML \qquad (5.9)$$

Any one of these steps may be rate-determining. For example, in the case of open-chain ligands such as **5.2**, the formation of the first M–N bond is found to be rate-determining, whereas with cyclam, it is the second M–N bond formation which determines the overall rate. The slower coordination of cyclam with respect to **5.2** can now be seen to originate in the reduced flexibility of the macrocyclic ligand. Consider the formation of the second M–N bond in each case. With **5.2**, the ligand can freely rotate around the C–C and C–N bonds to position the second nitrogen atom correctly for coordination without steric consequences for the remainder of the ligand. In contrast, for cyclam to position the nitrogen donor prior to coordination, the ligand will be forced into a high energy conformation, and the energy barrier for the reaction will be correspondingly higher (Fig. 5.2).

The conformational change within the ligand is important, and in the case of cryptands such as [2.2.2], it is thought that the conformation of the ligand changes at the stage of outer-sphere complex formation. In solution, the dominant form is the *endo, endo* conformation, but this has no outwardly oriented lone pairs for coordination, and it is only the minor *exo, endo* (or *exo, exo*) conformer that can lead to reaction (Scheme 5.1).

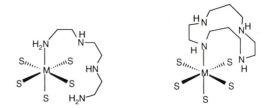

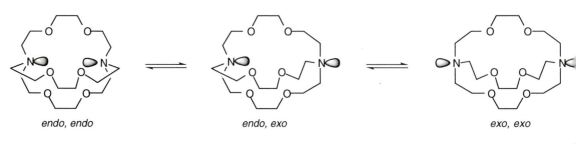

Fig. 5.2. Possible transition states prior to the formation of the second M–N bond for **5.2** and cyclam ligand coordination; S = solvent molecule.

endo, endo   *endo, exo*   *exo, exo*

Scheme 5

To summarize, the rate of formation of macrocyclic complexes is usually slower, and sometimes significantly slower, than that of open-chain ligands.

5.7 Dissociation of macrocyclic complexes

Equation 5.6 introduced the relationship between K, k_f and k_b. For tetraaza macrocycles, the high stabilities are due to extremely slow dissociation reactions with small k_b values. Typically, k_b values are 10^5 to 10^7 times smaller than those of open-chain analogues. However, a detailed study of the dissociation kinetics reveals a much more complicated situation. Complexes of tetraaza macrocyclic ligands are so stable that dissociation in neutral solution is unmeasurably slow. Accordingly, the measurements are usually made in acidic solution. The rate laws show varying dependence of $[H^+]$, indicating the involvement of intermediates in which one or more of the nitrogen donors are protonated – *i.e.* species in which one or more M-L bonds have been broken. This provides the mechanism by which dissociation occurs - a *stepwise* 'peeling off' of the donors rather than a catastrophic removal of all the donors at once.

   With crown ethers and cryptands, the stability is again seen to be due to k_b being smaller than for k_f; for example, for K^+ with [2.2.2] in water $k_b = 38$ s^{-1} and $k_f = 7.4 \times 10^6$ M^{-1} s^{-1}, whilst with 18-crown 6 in water $k_b = 3.7 \times 10^6$ s^{-1} and $k_f = 4.3 \times 10^8$ M^{-1} s^{-1}. The 'tighter fit' of the metal ion in the cryptate means that k_b is smaller than for the crown ether. If the hole-size is not optimal, the dissociation is easier, as seen by a comparison of the Li$^+$ complexes of [2.1.1] and [2.2.1], where k_b for the complex with [2.2.1] is

10^4 greater than for the [2.1.1] complex or for the Na$^+$ complexes with 18-crown-6, where $k_f = 2.2 \times 10^8$ M^{-1} s^{-1} is very similar to the formation rate of the K$^+$ complex whereas $k_b = 3.4 \times 10^7$ s^{-1} is considerably faster.

5.8 Hole-size effects and metal ion selectivity

Much of the early interest in macrocyclic ligands lay in the specificity that they appeared to show for particular metal ions. We have introduced the concept of hole-size in Chapter 2 and indicated that it provides a rule-of-thumb measure of suitability of a metal ion for a particular ligand, although we stressed repeatedly that conformational changes made the values very soft. It follows that a ligand in its minimum energy metal-binding conformation will be optimized for a particular size of metal ion and that when other metal ions are bound, the ligand conformational energy will rise with a resultant decrease in the stability of the complex. This is the origin of macrocyclic selectivity for metal ions. *It is very important to stress a hole-size metal ion mismatch does not usually mean that no complexation will occur, rather that the stability of the complex that is formed will be decreased.*

This effect was recognized very early in the development of macrocyclic chemistry and much attention concentrated upon the design of ligands for the selective binding of Group 1 and Group 2 metal ions. Figure 5.3 presents some data for Group 1 metal ions with 18-crown-6 and confirms that our subjective ideas about an optimal hole-size are borne out by the thermodynamic data. The data also illustrate another point – in general, for a given ionic size, the stability of the complex increases with increasing charge. In fact this latter observation is consistent with a picture of the bonding in these complexes which has a high degree of covalent character. This is additionally supported by a detailed consideration of the vast number of crown ether complexes which have been crystallographically characterized – although it is not possible with conventional X-ray experiments to 'see' the lone pairs, the geometry is often such that the oxygen-metal ion vector lies between the expected position of the lone pairs. This is consistent with a more electrostatic type of bonding.

Numerous functionalized crown ethers have been prepared in attempts to optimize the binding of particular metal ions. Much of the interest has centred upon the incorporation of additional functionality which can act as a spectator group for the binding of the metal ion – in other words, the ligands can be used as sensors for the detection of particular metal ions or as metal-ion triggered switches. A typical example is seen with the ligand **5.17** which has been used in ion-selective electrodes which are specific for lithium. Such ion-selective electrodes are of interest in the analysis of Group 1 metal ions in biological fluids. Selectivity in binding is also the basis of systems designed for ion transport across membranes. The basic model consists of a liquid or a polymer membrane containing a mobile crown ether or related compound. This can specifically bind a metal ion at one side of the membrane, diffuse through the membrane, and release the metal on the other side. This simple description refers to a passive process in which the direction of ion transfer is controlled by the metal ion concentrations on

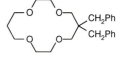

(5.17)

either side of the membrane. An example of a ligand used in this way is compound **5.18**, which has been used for the transport of potassium ions across a composite polymer/liquid crystal membrane. This latter area has offered some useful insights into the role of sodium and potassium ions in neurotransmission, although the sodium transport process is active – the sodium ions are *pumped* across the membrane.

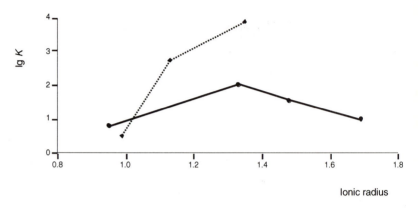

Fig. 5.3. Stability constant data (lg K) for complexes of Group 1 monocations (solid line) and Group 2 diactions (broken line) with 18-crown-6. Conventional ionic radii have been used.

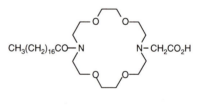

(5.18)

picrate

Considerable attention has been given to the design of ligands for the selective extraction of metal ions from aqueous solution into organic phases and most of the early studies were concerned with Group 1 metal ions. One of the difficulties with these studies is the role of the counterion. Numerous studies have used alkali metal picrates because the yellow colour of the extracted picrate in the organic phase allows a rapid quantification of the extraction efficiency by spectrophotometry. However, numerous crystallographic studies have established that picrate ions are usually coordinated to the metal ion in Group 1 metal picrate-crown ether complexes. Thus, the extraction ratios and thermodynamic parameters determined refer *only* to the system in which the counter-ion is picrate and cannot be simply extended to predict the behaviour when mixtures of simpler anions are present.

The presence of mixed donor atoms within the macrocyclic ligand may have an effect on the stability even if the ring size remains constant. This is emphasized in the stability of the potassium complexes (in methanol) with the series of ligands **5.19-5.22** (Fig. 5.4). Part of the effect arises from the variation in hole-size as the donors vary, but a part comes from the preference of the hard potassium ion for hard donor atoms.

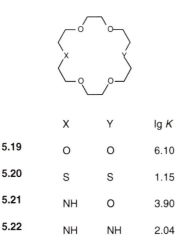

	X	Y	lg K
5.19	O	O	6.10
5.20	S	S	1.15
5.21	NH	O	3.90
5.22	NH	NH	2.04

Fig. 5.4. The stability of potassium complexes with a series of ligands related to 18-crown-6 (in methanol solution).

The selective extraction of transition metal ions is a topic of enormous commercial interest. The traditional methods of obtaining pure metals from their ores are based upon pyrometallurgy with thermal processes such as smelting. There is increasing concern at the environmental impact of these processes which release large amounts of environmentally damaging gases into the environment. Reduction of metal oxides by carbon results in the release of toxic carbon monoxide or the greenhouse gas carbon dioxide. Metals that occur as sulfide ores will give sulfur dioxide, one of the principle gases involved in the *acid*-rain phenomenon, upon thermal processing in air.

Accordingly, new methods of *hydrometallurgy* are being investigated in which as much of the chemistry and isolation as possible is performed in aqueous medium. An added advantage is that it is possible to work at much lower concentrations of metal and it is profitable to rework spoil heaps or ore deposits not previously though to be viable.

The basic methodology is relatively simple. A solution containing the metal ion of interest is treated with a solution of the extractant (ligand) in an immiscible organic solvent such as paraffin. The metal ion is extracted from the aqueous phase to the organic phase which may then be separated. Back-extraction of the metal ion from the organic phase into a new aqueous phase gives a solution enriched in the metal ion of interest (or, in perfect cases, containing only the ion of interest). The back-extraction is often achieved by the use of an aqueous solution which contains different chloride ion concentrations or is at a different pH to the original phase. These are known as *chloride-swing* or p*H-swing* methods. Finally, the metal ion is concerted into the desired form by conventional methodology. For example, in the case of copper extraction, the final result is an aqueous solution containing copper(II) salts. Electrolysis yields extremely pure copper metal.

The problem that all hydrometallurgists are working against is the *Irving-Williams* series. This is an experimentally observed series which finds that for the majority of ligands the stability of their complexes with first row transition metals is given by:

The Irving-Williams series

$$K(Mn^{2+}) < K(Fe^{2+}) < K(Co^{2+}) < K(Ni^{2+}) < K(Cu^{2+}) > K(Zn^{2+})$$

Whilst this is good news if the object is to extract copper, there are obviously difficulties if the aim is to selectively bind nickel(II) in the presence of copper(II). The Irving-Williams series originates in a combination of ligand-field effects (where the ligand field stabilization energies are expected to be at a maximum for high-spin nickel(II) complexes) and the Jahn-Teller distortion observed in six-coordinate complexes of the d^9 copper(II) ion. It was hoped that hole-size effects would provide an additional factor allowing selectivity to be tuned and for maximal binding to occur at metal ions other than copper(II).

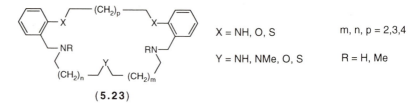

X = NH, O, S	m, n, p = 2,3,4
Y = NH, NMe, O, S	R = H, Me

(**5.23**)

As a case history, the series of ligands of general structure **5.23** are of interest. Although tetraaza macrocycles might appear to be the obvious choice for designing selective ligands for transition metal ions, they have a number of disadvantages. The most important problem lies in the kinetic properties that we have discussed earlier in this chapter. We have seen that both the complexation and decomplexation reactions are slow. In an ideal extraction process, the equilibrium distribution of metal ions between the two phases should be achieved rapidly - in particular, the back-reaction, involving decomplexation of the macrocyclic complex should be fast. This is particularly important if the selectivity for the desired metal ion is not too high and the extraction-back-extraction sequence needs to be repeated a number of times to achieve efficient separation. Accordingly, the mixed donor macrocycles **5.23**, which have reasonably rapid forward and back complexation reactions, have many advantages.

The copper(II) complexes with these ligands have stabilities varying over 10 orders of magnitude, with the maximum destabilization (lowest stability) occurring when sulfur donors are incorporated. This fits with our subjective feeling that copper(II) is hard. However, a similar trend is seen in the

	X	Y
5.24	O	NH
5.25	NH	NH
5.26	O	O
5.27	NH	O
5.28	S	NH
5.29	O	S
5.30	S	O
5.31	S	S

stability of the nickel(II) complexes and no inversion of the Irving-Williams order occurs.

The true value of series of ligands of this type is seen in the maximizing of selectivity for another pair of metal ions, silver(I) and lead(II). One way of expressing the efficiency of a particular ligand is by giving the Δlg K values (Δlg K is the difference in the stability of the silver and the lead complexes with the macrocyclic ligand). The larger the value of Δlg K, the greater the selectivity for the silver ion. Data for a series of complexes are given in Fig. 5.5.

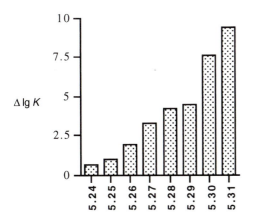

Fig. 5.5. Δlg K values for silver(I) and lead(II) complexes of the ligands **5.24-5.31**. In each case the silver complex is the more stable.

To conclude this section, it is perhaps relevant to ask if it is really necessary to use macrocyclic ligands to optimize extraction selectivity. Obviously the series of ligands discussed above are capable of optimization, but throughout this book we have noted that the synthesis of macrocyclic ligands is not always easy and that special methods of ring-closure often have to be used. If these ligands are to be successfully used for hydrometallurgy, it is necessary to prepare multi-kilogram quantities. And even in the most efficient recycling process, some ligand loss will occur between extraction cycles.

The majority of hydrometallurgical processes involving the extraction of copper utilize ligands such as **5.32**, which is clearly not a macrocycle! In fact, **5.32** may be thought of as a *pre-macrocyclic ligand* and the principles embodied in this description offer intriguing new possibilities for ligand design. Although **5.32** is not a macrocycle, two molecules can hydrogen bond to form the *pseudomacrocyclic* system **5.33**. This is really just a macrocyclic ligand in which two of the peripheral covalent bonds have been replaced by hydrogen bonds. The dimer behaves as a N_2O_2 donor to give the copper complex **5.34**. The substituents R_1 and R_2 are long chain alkyl groups which confer solubility in non-aqueous solvents and the copper complex transfers to the non-aqueous phase. The formation of the copper complex **5.34** is accompanied by ligand deprotonation. This allows the copper to be back-extracted from the complex by a pH-swing; extraction into acidic aqueous solution protonates the ligand and releases copper(II) ions (Fig. 5.6).

(5.32)

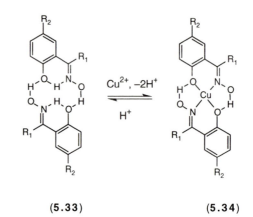

Cu^{2+}, -2H$^+$

H$^+$

(5.33)   **(5.34)**

Fig. 5.6. The extraction of copper(II) by a dimeric *pseudomacrocyclic* ligand is controlled by a pH-swing. The hydrogen bonds are not explicitly distinguished from other covalent bonds within the structures.

The observation that hydrogen-bonding (and, indeed, other secondary interactions) may be used to control macrocycle formation is likely to lead to numerous new classes of ligands with unusual properties. We conclude this chapter with a question – should the nickel(II) complex of dimethylglyoxime **5.35**, familiar to generations of students as a red precipitate obtained in the gravimetric analysis of nickel, be regarded as a pseudomacrocyclic complex in which the ligand is a dimer of the oxime?

(5.35)

6 Reactivity of macrocyclic complexes

Introduction

It would be impossible to discuss the reactivity of macrocyclic complexes in detail in a book many times this length, let alone within one Chapter. However, I hope that this Chapter will leave the reader with a taste for the various types of reaction which are possible.

6.1 Axial substitution reactions

One of the simplest reactions which could be envisaged is the exchange of axial ligands X and Y in a macrocyclic complex $[ML(X)(Y)]^{n+}$ (Fig. 6.1). The high kinetic and thermal stability of macrocyclic complexes means that it is possible to design systems in which no competitive displacement of the macrocyclic ligand occurs. Even with chelating open-chain ligands such high specificity is hard to find. As a result, macrocyclic complexes have been used for a number of fundamental investigations into the mechanism of substitution of ligands in transition metal complexes. The majority of studies have involved cobalt(III) complexes of tetraaza macrocyclic ligands. By using large numbers of complexes with different X and Y axial ligands and with different incoming ligands it is possible to probe very precisely such things as the *trans effect* in octahedral complexes. Another advantage of such systems is that there is usually no stereochemical change in the course of the reaction.

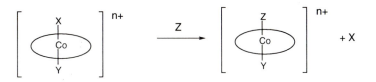

Fig. 6.1. The displacement of axial ligand X by Z in a cobalt(III) complex. Studies of such reactions have given valuable insight into the mechanisms of substitution reactions at transition metal centres.

Much of the interest in transition metal complexes of porphyrins and related ligands is coupled with the important biological roles that these compounds play. Naturally, numerous synthetic porphyrins have been prepared which are intended to model such biological functions as the binding of dioxygen. This is, of course, just a special case of a ligand displacement reaction, where the incoming ligand is a dioxygen molecule and the leaving ligand is a solvent (usually water).

Accordingly, there has also been much interest in the synthesis of simpler, non-porphyrin macrocycles which also give complexes which can bind dioxygen reversibly. Surprisingly large numbers of compounds are now known which bind oxygen with varying degrees of efficiency and varying degrees of reversibility. Although complexes with open-chain ligands are of increasing importance, many of the earliest and still the most efficient, systems are based upon macrocycles.

As discussed earlier, the problem is in avoiding the irreversible formation of μ-oxo complexes. One of the approaches that has been developed involves the linkage of two macrocycles together such that dioxygen can generate a bridging peroxo or superoxo ligand between two metal centres, but which are sufficiently rigidly linked that the metal centres cannot close up to permit the formation of a μ-oxo bridge. Cobalt(II) and iron(II) complexes of ligands such as **6.1** have been found to reversibly bind dioxygen under specific conditions.

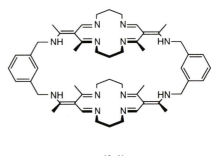

(6.1)

Related mononuclear systems have also been prepared and shown to bind dioxygen. In these case, the trick is to place a bulky strap over the top of the macrocycle such that dimerization to give μ-oxo compounds is not possible. One such example is the ligand **6.2**, which forms an iron(II) complex which can bind dioxygen. Macrocycles of this type have been called *lacunar* ligands.

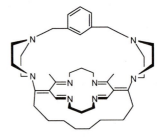

(6.2)

6.2 Reactions of the coordinated macrocyclic ligand

The high stability of macrocyclic complexes means that it is often possible to perform reactions upon the coordinated ligand which would lead to decomposition of the complex if open-chain ligands were used. These reactions may be used for the subsequent structural development of the macrocyclic ligand to incorporate additional functionality or to develop three-dimensional cage structures from two-dimensional cores.

One of the very simplest reactions that might be envisaged is the alkylation of a nitrogen donor, to convert a secondary amine to a tertiary amine. If this reaction is attempted on the free amines, it is not usually possible to halt the reaction at the monoalkylation stage and significant amounts of quaternized ammonium salts are obtained. In a coordinated amine, the metal acts as a protecting group and prevents the formation of the quaternized compound. A typical example of an alkylation reaction is seen in the methylation of the complex $[Ni(cyclam)]^{2+}$ with methyl iodide to give $[Ni(Me_4cyclam)]^{2+}$. The interest in this reaction lies in the fact that the product is not the same as that obtained from the reaction of nickel(II) salts with $Me_4cyclam$. The direct reaction with $Me_4cyclam$ gives a complex with the conformation in Fig. 6.2 a (which has *RSRS* absolute stereochemistry at the nitrogen centres) whereas the alkylation of $[Ni(cyclam)]^{2+}$ gives the favoured *RSSR* diastereomer seen in Fig. 6.2 b (see also p. 29). The methyl substituents have a significant steric influence and $Me_4cyclam$ tends to form five coordinate complexes with a single axial ligand (Fig. 6.2).

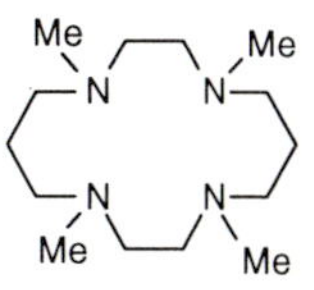

Me₄cyclam

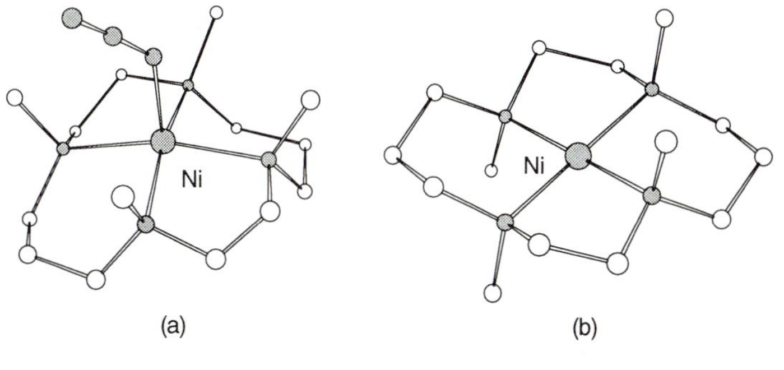

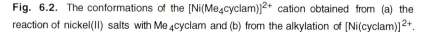

(a)   (b)

Fig. 6.2. The conformations of the $[Ni(Me_4cyclam)]^{2+}$ cation obtained from (a) the reaction of nickel(II) salts with Me₄cyclam and (b) from the alkylation of $[Ni(cyclam)]^{2+}$.

Perhaps more interesting are reactions of the carbon framework of the macrocyclic ligand with electrophiles. Numerous examples of such reactions are known, many of which are based upon macrocycles such as the ligand in **6.3** which contains the diaza analogue of a β-diketonate. A typical reaction is shown in Fig. 6.3 in which the nickel(II) complex **6.3** is methylated at the nucleophilic carbon of the ligand. The ligand in **6.3** has been explicitly drawn in the deprotonated form, and it is the neutral nickel(II) complex of the double-deprotonated ligand that is usually isolated. The closely related complex **6.4** deprotonates readily to give **6.5** and has been shown to have a

pK$_a$ of 6.45 which indicates that the methylene protons are significantly activated upon coordination.

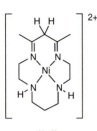

(6.4)

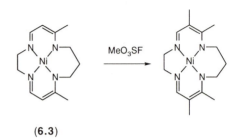

(6.3)

Fig. 6.3. The alkylation of a nickel(II) complex of a doubly-deprotonated ligand occurs at carbon.

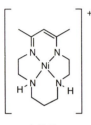

(6.5)

Exploiting the reactivity of these nucleophilic carbon sites in macrocyclic ligands opens up an entire range of chemistry. Reactions are not limited to the use of simple alkylating agents and the reaction of **6.6** with ethyl acrylate leads to the *C*-functionalized complex **6.7**. In contrast, when cyclam or [Ni(cyclam)]$^{2+}$ are reacted with ethyl acrylate or acrylonitrile, reaction occurs at the amines to give mixtures of *N*-functionalized compounds.

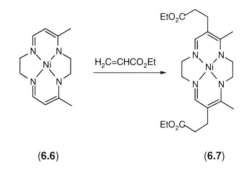

(6.6)   **(6.7)**

It is even possible to acylate macrocyclic complexes of this type and these reactions give an entry into compounds such as **6.1** and **6.2**. The acylation of **6.6** with MeCOCl gives the doubly acylated complex **6.8** in good yield. If this new complex is now reacted with strong alkylating reagents such as MeSO$_3$F, a new complex containing two enol ether functionalities **6.9** is obtained. This compound is electrophilic and reaction with amines results in the displacement of methanol and the formation of a *bis*(enamine) **6.10**. It is compounds such as **6.9** and **6.10** which are important in the capping reactions to give **6.1** and **6.2**. A typical example is seen in the preparation of the capped macrocyclic complex **6.11** which maybe obtained from the reaction of **6.9** with octane-1,8-diamine or **6.10** with 1,8-dibromooctane (Fig. 6.4). The ability of the iron(II) and cobalt(II) complexes of these ligands to bind dioxygen has been mentioned earlier. Complex **6.2** is formed from the sequential reaction of **6.9** with piperazine followed by 1,3-bis(bromomethyl)benzene.

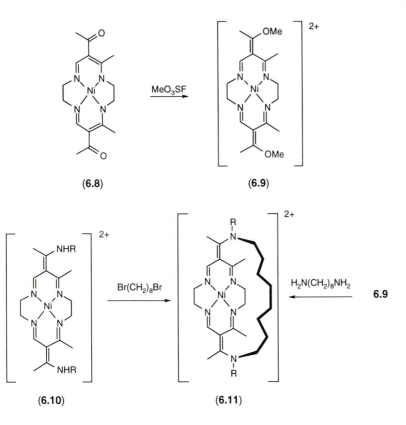

(6.8)   **(6.9)**

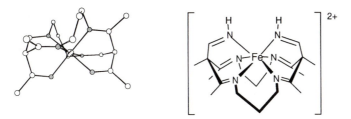

(6.10)   **(6.11)**

Fig. 6.4. The reactions of enol ether derivatized macrocyclic complexes with α, ω-diamines or of enamine-functionalized compounds with α, ω-dihalides allows the preparation of capped *lacunar* systems.

It is not even necessary for an external electrophile to be involved, and in extreme cases additional ligands bound to the metal centre may react with the macrocyclic framework. A good example of this is seen in the reaction of the complex $[Fe(\textbf{6.12})(MeCN)_2]^{2+}$ with a mild base such as triethylamine. Deprotonation of the methylene groups yields nucleophilic carbon centres which undergo an intramolecular reaction with the acetonitrile ligands to generate **6.13** which contains a six-coordinate octahedral iron(II) centre. Interestingly, the solid state structure of the complex reveals that the final structure has a *cis* arrangement of the functionality derived from the original acetonitrile ligands.

(6.12)

(6.13)

Finally, it must be noted that it is possible to perform conventional 'organic' electrophilic substitution reactions upon aromatic rings if the macrocyclic complexes are sufficiently stable. A typical example is seen in the bromination of the aromatic ring of **6.14** upon reaction of bromine (Fig. 6.5). The reaction is thought to proceed by an initial *N*-bromination followed by an intramolecular bromine atom transfer.

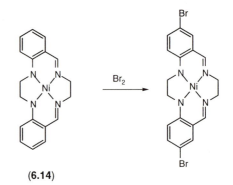

(**6.14**)

Fig. 6.5. In suitable cases, it is possible to perform electrophilic substitution reactions on the aromatic rings of macrocyclic ligands.

In addition to reaction of the ligand with electrophiles, unsaturated macrocyclic complexes can also react with nucleophiles. Some of the best examples are found in the reactions of imine ligands. We have previously mentioned that imine formation is a stepwise process and that aminols are involved as intermediates:

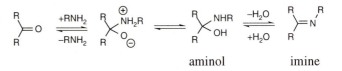

aminol   imine

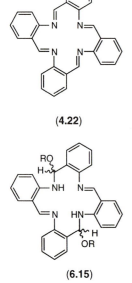

(**4.22**)

(**6.15**)

The sp^3 hybridization at the central carbon in the aminol and the sp^2 hybridization in the imine mean that the conformations associated with the two compounds will differ and we have seen examples of this in ring contraction reactions. In particular, the aminol will be non-planar whereas the imine can participate in planar conjugated structures. The change from the imine to the aminol results in an effective reduction in hole size. Many examples are known in which reactions expected to give imine macrocycles give ligands in which an aminol is present. In other cases, imine ligands react with water or alcohols to generate sp^3 hybridized centres. For example, nickel(II) ions are slightly too large for the cavity in **4.22** and reaction of the [Ni(**4.22**)]$^{2+}$ complex with alcohols in basic conditions generates complexes of **6.15** in which the addition of two alcohol molecules to the imine bonds results in the formation of a smaller, non-planar bonding cavity. A mixture of the various diastereomers is expected.

In some cases these reactions may be synthetically useful in the structural development of a macrocyclic ligand. The complex [Ni(**4.22**)]$^{2+}$ is activated to attack by nucleophiles as a results of the hole-size of the ligand being slightly too large for the metal ion. If a diol is reacted with [Ni(**4.22**)]$^{2+}$ it should be possible to use the reaction for the preparation of capped systems. A typical example of such a reaction is presented in Fig. 6.6 in which the diol incorporates additional axial donor capacity. In this case, the stereochemistry at the two new sp^3 hybridized centres is defined, since the capping group must lie above a face of the macrocycle.

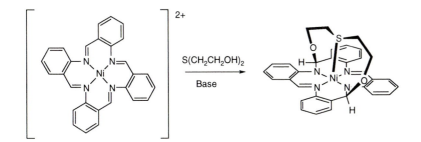

Fig. 6.6. The addition of nucleophiles to imines may be used for the structural development of macrocyclic ligands.

6.3 Demetallation and metal exchange reactions

When template reactions were discussed in Chapter 4, it was mentioned that it was found that cyclization reactions were very often template metal ion specific. It is sometimes possible to demetallate the template products to give free ligands which may then be coordinated to the metal of choice. This approach is limited to ligands which are stable and also to those which do not undergo subsequent reactions under the demetallation conditions. Furthermore, as we saw in Chapter 4, hole-size effects may result in rearrangement in the second coordination step.

Because of the high kinetic and thermodynamic stability of macrocyclic complexes (see Chapter 5) it is necessary to use forcing conditions to remove metal ions from the complexes. For example, to demetallate [Ni(cyclam)]$^{2+}$ complexes it is necessary to use concentrated cyanide solutions. Similarly, to remove the copper(I) centres from catenates, treatment with cyanide is required. This latter reaction leaves the metal-free interlinked rings. In the case of the cobalt(III) sepulchrates formed by the template condensation of [Co(en)$_3$]$^{3+}$ with formaldehyde and ammonia, it has proved to be impossible to remove the metal ion from the ligand. Unsaturated ligands, in particular those containing imine bonds, may react with good nucleophiles such as cyanide.

In some cases it is possible to remove metal ions under relatively mild conditions. This is often the case with complexes of Group 1, 2 or 12 metal ions or in cases where the metal ion is poorly matched to the ligand binding

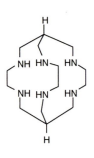

(2.27)

cavity. Although it is impossible to remove cobalt from cobalt-sepulchrate complexes, mercury(II) complexes of the related cage ligand **2.27** may be demetallated by treatment with chloride. In this case, the mercury is removed as the $[HgCl_4]^{2-}$ ion. Phthalocyanine may be prepared from the dilithium complex Li[Li(pc)] by reaction with water or dilute acid. As the dilithium compound is very conveniently prepared from 1,2-dicyanobenzene this represents a useful synthesis of the metal-free ligand. Another example where chloride ion is used to remove a metal ion from a macrocyclic complex is found in the demetallation of $[Ni(\mathbf{6.16})]^{2+}$ – the nickel(II) ion is too small for the cavity and it is thus labilized and may be removed as $[NiCl_4]^{2-}$ upon treatment with HCl.

In exceptional cases, template reactions result in the formation of free ligands. Although nickel(II) acts a template for the formation of $[Ni(\mathbf{2.7})(H_2O)_2]^{2+}$ from the bis(hydrazine) and glyoxal, other transition metal cations are ineffective. However, $[H_2(\mathbf{2.7})]^{2+}$ is obtained from the template condensation in the presence of $SnMe_2Cl_2$; presumably as a result of a hole-size mismatch between the macrocycle and the dimethyltin(IV) cation, the initial product is labile and deposits tin(IV) oxide. Subsequent reaction of the free ligand with transition metal or even Group 1 metal ions results in the formation of the desired complexes of **2.7**.

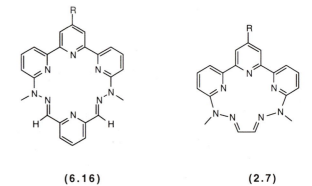

(6.16)   **(2.7)**

One other approach which has been adopted involves the *exchange* of one metal ion for another without the formal isolation of the metal–free ligand. Zinc, cadmium and mercury complexes have all been used in this way, as have complexes with Group 2 and Group 1 metal ions. One of the problems with phthalocyanine chemistry is the insolubility of the free ligand in almost all solvents. In contrast, the dilithium derivative Li[Li(pc)] is soluble in solvents such as acetone and is a valuable intermediate for the synthesis of other phthalocyanine complexes:

$$Li[Li(pc)] + ZnCl_2 = 2LiCl + [Zn(pc)]$$

In earlier chapters a number of examples of macrocyclic ligands obtained from the template condensation of α,ω-dicarbonyl compounds with α,ω-diamines have been described. Once again, these reactions are rather sensitive to the template metal ion and not all transition metal ions are effective.

However, metal ions such as lead(II) and barium(II) have been found to be effective in the formation of the [2+2] products and subsequent treatment with first row transition metal ions results in transmetallation processes, although in many cases rearrangement of the ligand also occurs. Such an example is the condensation of **6.17** with 2,6-diacetylpyridine in the presence of barium(II) gives a barium complex of **6.18**. This latter compound itself is of some interest; the Ba^{2+} ion is too small for the cavity of the [2+2] macrocycle and a ring-contraction occurs by the addition of the central secondary amine to an imine bond to give an 18-membered ring macrocycle. Subsequent reaction with copper(II) salts results in a reversal of this ring contraction to give **6.19** as a dinuclear $[Cu_2(\mathbf{6.19})]^{2+}$ complex.

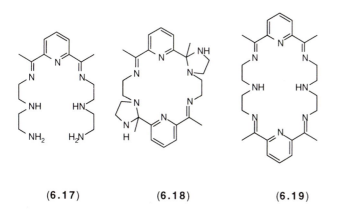

(6.17)   (6.18)   (6.19)

6.4 Metal-centred oxidation and reduction reactions

Macrocyclic ligands have proved capable of stabilizing a wide range of 'unusual' oxidation states and redox reactions involving macrocyclic complexes have been studied in considerable detail. It is convenient to consider redox reactions in which the formal oxidation state of the metal changes and those in which the structure of the ligand changes.

Saturated ligands such as cyclam, triazacyclonane (tacn) or trithiacyclononane (**1.7**) provide an unambiguous starting point. Simple electron transfer reactions will be metal-centred and the formal oxidation state will correspond to the 'real' oxidation state of the metal. The sandwich complex $[Pd(\mathbf{1.7})_2]^{2+}$ may be oxidized electrochemically to the unusual palladium(III) complex $[Pd(\mathbf{1.7})_2]^{3+}$. This represents one of the surprises of working with such ligands as the soft donor set is expected to stabilize lower oxidation states. The platinum analogue undergoes a similar oxidation to a rare mononuclear platinum(III) species. The silver(I) species $[Ag(\mathbf{1.7})_2]^+$ may also be oxidized to a silver(II) compound, $[Ag(\mathbf{1.7})_2]^{2+}$ although in this case the compound is unstable much above $-70°C$.

(1.7)

E° [Ni(cyclam)]$^{3+/2+}$ = +0.67 V *versus* Ag$^+$/Ag in MeCN.

E° [Ni(**2.42**)]$^{3+/2+}$ = +1.3 V *versus* Ag$^+$/Ag in MeCN.

2[Ag(L)]$^+$ = Ag + L + [Ag(L)]$^{2+}$

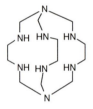

(2.25)

It was recognized very early on that tetraaza macrocyclic ligands were capable of stabilizing nickel(III) and copper(III) oxidation states and the metal(III) complexes are readily accessible by chemical or electrochemical oxidation of the metal(II) species. The exact redox potential is very dependent upon the structure of the ligand with nickel(II)/(III) potentials ranging over a span of about 2 V for structurally related tetraaza macrocyclic ligands. There are the expected hole-size effects although it is not easy to rationalize the variation – nickel(III) is less favoured in both 13- and 15-membered rings than in cyclam, even though nickel(III) is a smaller ion than nickel(II). In general, anionic macrocyclic ligands stabilize higher oxidation states better than neutral ligands.

Cobalt complexes of the tetraaza macrocycles show the expected cobalt(II)/(III) redox processes and the complexes [Co(L)X$_2$]$^{n+}$ show a marked dependence upon the axial ligands. Both outer sphere and inner sphere electron transfer reactions have been observed with cobalt complexes. The cobalt(II)/(III) potential is also very dependent upon the hole-size of the ligand, representing the differing ionic radii of the two oxidation states.

Cyclam can support a range of unusual oxidation states including silver(II) and even silver(III). In fact, silver(II) is so stabilized that silver(I) disproportionates to silver(II) and silver metal in the presence of tetraaza macrocyclic ligands.

It is also possible to observe metal-centred reductions to give unusual oxidation states using these saturated macrocyclic ligands. For example, in addition to showing a nickel(II)/(III) process, [Ni(cyclam)]$^{2+}$ may also be reduced to a nickel(I) complex at an accessible (–1.7 V *versus* Ag$^+$/Ag) potential. As expected, anionic ligands destabilize the nickel(I) state.

The more rigid ligands and the encapsulating ligands are also interesting. The cobalt(III) complex of **2.25** is obtained directly from the reaction of [Co(en)$_3$]$^{3+}$ with formaldehyde and ammonia. With 'normal' ligands, cobalt(III) complexes are kinetically inert, but cobalt(II) is labile and in aqueous solution, reduction of cobalt(III) compounds is expected to lead to the formation of [Co(H$_2$O)$_6$]$^{2+}$. In the case of the sepulchrates, the cobalt(II) ion is trapped within the cage and the reduction product is [Co(**2.25**)]$^{2+}$, which undergoes *no* metal ion exchange of coordinated cobalt(II) with [Co(H$_2$O)$_6$]$^{2+}$. A number of other unusual oxidation states, such as silver(II), have been stabilized within sepulchrate and related ligands. The ability of crown ethers and cryptands to react with alkali metals to give electrides or sodides has been commented upon earlier.

Perhaps even more remarkable is the behaviour of copper(I) catenates in which a copper(I) ion is bound to two interlocked rings (see Chapter 4). In some of these compounds a reversible copper(I)/(0) process is observed to give a stable copper(0) catenate which does not eliminate copper metal!

When unsaturated ligands are involved the discussion becomes a little more complicated and it is necessary to distinguish between 'formal' and 'actual' oxidation states. A good example is provided by zinc(II) porphyrin complexes. These compounds exhibit two oxidation and two reduction processes corresponding to formal oxidation states of +III, +IV, +I and 0 respectively. As the known chemistry of zinc is almost exclusively that of the divalent state, these assignments seem unlikely. The problem lies in the unsaturation within the ligand. The porphyrin ligand (H_2L) has a delocalized π-bonding system with both filled π- and low-lying empty π^*-orbitals. The oxidation processes correspond to the removal of electrons from the ligand π-orbitals to give ligand cations L^+ and L^{2+} rather than from the metal orbitals. Similarly, the reductions involve the placing of electrons into the ligand π^*-orbitals to give radical anions L^- and L^{2-}. The difficulty comes when the zinc ion is replaced by a metal ion such as iron which can also undergo redox reactions. The problem comes in correlating the observed reduction potentials to particular electronic arrangements. For example, does the first oxidation process of an iron(II) porphyrin complex correspond to the formation of an iron(III) complex $[Fe^{III}(L)]^+$ or an iron(II) ligand cation $[Fe^{II}(L^+)]$? It is necessary to perform ESR or Mössbauer spectroscopic studies in order to determine whether the unpaired electrons are localized on the metal centre or the ligand and in many cases the answers are not unambiguous. It is certainly the case that iron(I), iron(II), iron(III), iron(IV) and possibly iron(V) and iron(VI) can exist within porphyrin and related ligands. As the understanding of the electronic subtlety of porphyrin-iron complexes is crucial to an understanding of the functioning of such compounds within biological systems it is unfortunate that so many ambiguities exist. The problem is exacerbated when the dioxygen adducts of these systems are considered – there are now additional ambiguities in the electronic state of the dioxygen (singlet neutral dioxygen ligand, triplet neutral dioxygen ligand, superoxide or peroxide).

The problem is easily understood. Conventional wisdom says that to stabilize low oxidation state transition metal complexes, π-acceptor ligands with low-lying π^*-orbitals should be used. Overlapping of metal orbitals with the ligand π^*-orbitals allows the back-donation of electron density to the ligand, thus stabilizing a low oxidation state metal centre. However, if the ligand is a very good π-acceptor, then it is possible that the π^*-levels will lie below the metal d orbitals. If this is the case, then reduction will involve the addition of electrons to the ligand π^* orbitals rather than the metal d orbitals.

In many cases, the design of macrocyclic ligands has optimized the stabilization of low oxidation states to such an extent that an ambiguity exists. Studies of the metal complexes of planar pentadentate macrocyclic ligands such as **2.7** or **6.20** reveal that the effects are very subtle. In some cases the nickel(II) complexes are reduced to $[Ni^I(L)]$ species, in other cases to nickel(II) $[Ni^{II}(L^-)]$ complexes. The localization of the electron may be altered by using electron-releasing or electron-accepting axial ligands in addition to making changes in the structure of the macrocyclic ligand itself.

In a final twist to this story, let us return to redox inactive metal ions. With saturated ligands, we saw that the most likely site for redox reaction was at a metal ion. We also saw that with metal ions such as zinc(II) which only have one common oxidation state (other than zero), ligand-centred processes were possible with ligands such as porphyrins. What happens with saturated ligands and redox inactive metal centres? Actually, we have already seen one of the consequences of this combination in the formation of sodide

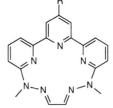

(**2.7**)

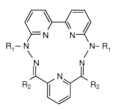

(**6.20**)

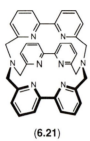

(6.21)

or electride complexes from the reaction of Group 1 metals with crown ethers or cryptands. In a final example, we consider ligand **6.21** which is an analogue of a cryptand which contains six nitrogen donors and which is a good π-acceptor ligand as it contains three 2,2'-bipyridine units. This ligand forms a normal complex cation [Na(**6.21**)]$^+$ with sodium bromide in which the sodium cation is within the cryptand. Electrochemical reduction leads to the sequential formation of [Na(**6.21**)], [Na(**6.21**)]$^-$ and [Na(**6.21**)]$^{2-}$. The cavity in the ligand is not large enough to contain a neutral sodium atom or a sodide anion and the crystal structural analysis indicates that the compounds should not be formulated as electrides. In these compounds, the electrons are localized on the 2,2'-bipyridine ligands and a more appropriate formulation of [Na(**6.21**)]$^{2-}$ would be [Na$^+$(**6.21**$^{3-}$)].

The discussion above should indicate that it many cases it is not really meaningful to describe an oxidation or reduction process as being metal- or ligand-centred. However, in the next section we will encounter redox reactions which are unambiguously ligand-centred in that they involve a change in the structure of the ligand as opposed to the placing of electrons in ligand π*-orbitals or the removal of electrons from π-orbitals.

6.5 Ligand centred redox processes

To date we have considered redox reactions in terms of the addition or loss of electrons – it is also convenient to think in terms of the addition or loss of hydrogen or oxygen atoms.

Numerous examples are known in which macrocyclic ligands undergo this type of redox process. Direct hydrogenation reactions of both free ligands and metal complexes are possible, although reactions of the free ligands are perhaps more common. The Me$_6$cyclam ligand **2.42** is obtained by the catalytic hydrogenation of the Curtis macrocycle **4.7**, whilst hydrogenation of dibenzo-18-crown-6 gives dicyclohexano-18-crown-6 **6.22**. It is, however, possible to reduce [Ni(**4.7**)]$^{2+}$ directly to [Ni(**2.42**)]$^{2+}$ using dihydrogen and a platinum catalyst. The reduction of the double bonds introduces new stereocentres and in many cases mixtures of the possible diastereomers are obtained.

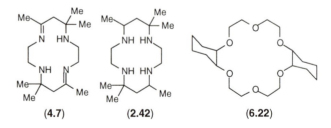

(4.7)   (2.42)   (6.22)

Oxidation of a coordinated macrocyclic ligand is a more commonly observed process and it is convenient to distinguish between oxidation processes which involve the addition of oxygen atoms and those which involve the loss of hydrogen atoms.

Many macrocyclic ligands have been designed for the preparation of iron, cobalt and nickel complexes which might bind dioxygen. We have already mentioned that one of the problems with such model complexes is the formation of dinuclear μ-oxo compounds. Another reaction pathway which can cause difficulties is oxygen atom transfer to the coordinated ligand.

The cobalt(III) complex of **6.23** provides a good example of this as it is not likely that dioxygen can enter the cage to interact with the metal ion directly. Oxidation of [Co(**6.23**)]$^{3+}$ with dioxygen in the presence of activated charcoal as a catalyst results in reactions of the encapsulating ligand. The first step is a dehydrogenation to form an imine **6.24** and subsequent reaction involves oxygen atom incorporation to form a bis(amide) complex **6.25** (Fig. 6.7).

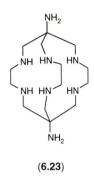

(**6.23**)

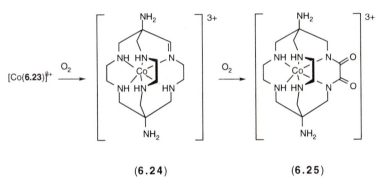

(**6.24**)   (**6.25**)

Fig. 6.7. The oxidation of a cobalt(III) complex of an encapsulating ligand leads sequentially to an imine and an amide.

The formation of imines is a common consequence of reaction with oxygen and numerous dehydrogenation products of macrocyclic complexes have been prepared deliberately or through serendipity. The macrocyclic complex **6.26** is obtained from the template condensation of the bis(hydrazone) of diacetyl, MeC(=NNH$_2$)C(Me)=NNH$_2$, with formaldehyde. Reaction of **6.26** with air leads to the fully conjugated complex **6.27** containing a new dianionic ligand. The driving force for the reaction would seem to be the formation of the aromatic 18-membered ring. A consequence is a 'rotation' of the ligand such that different nitrogen donor atoms are used in **6.27** to the starting material **6.26**.

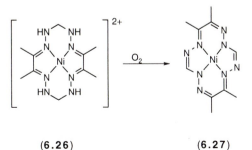

(**6.26**)   (**6.27**)

A range of other oxidizing agents may be used and when the scope of these reactions is considered it is seen that the rationalization in terms of conjugation above was perhaps naive. For example, the iron(II) complex $[Fe(\mathbf{4.7})]^{2+}$ is oxidized to $[Fe(\mathbf{6.28})]^{2+}$ upon aerial oxidation. In this case, there has been no rotation of the ligand, but the two original imine double bonds have migrated. In contrast, the oxidation of $[Ni(\mathbf{4.7})]^{2+}$ by nitric acid leads sequentially to the non-conjugated compounds $[Ni(\mathbf{6.29})]^{2+}$ and $[Ni(\mathbf{6.30})]^{2+}$.

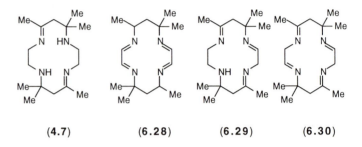

(4.7)   (6.28)   (6.29)   (6.30)

Hydride or hydrogen atom abstraction agents may also be used with macrocyclic complexes and a typical reaction involving the trityl cation, Ph_3C^+ is presented in Fig. 6.8. Note that in this case, the nickel complex does lead to a conjugated system after oxidation.

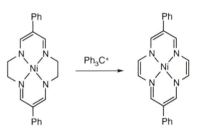

Fig. 6.8. Oxidation of a nickel(II) complex can lead to conjugated systems.

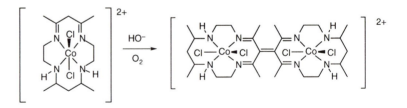

Fig. 6.9. Oxidative dimerization of a cobalt(III) complex.

A final possible outcome of an oxidation reaction is the dimerization of two macrocycles to give a new dinucleating system and an example of this is given in Fig. 6.9. The starting cobalt(III) complex reacts with dioxygen in alkaline conditions to form a new dicobalt(III) complex. The first step of the reaction is undoubtedly deprotonation of the active methylene group with a subsequent dimerization to a dihydro compound which is dehydrogenated to give the observed product. Remarkably, the axial ligands are retained in this reaction.

7 Other aspects of macrocyclic chemistry

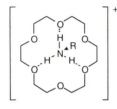

Fig. 7.1. The binding of an ammonium ion to 18-crown-6.

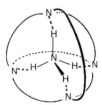

Fig. 7.2. The binding of **7.3** to an ammonium ion.

7.1 Binding of organic cations

To date, we have assumed that coordination chemistry is limited to inorganic cations. However, we have also stated that the interaction of Group 1 metal ions with crown ethers and cryptands has a significant electrostatic component and this suggests that organic cations such as ammonium ions might also be bound by macrocyclic ligands.

With ions such as $[NH_4]^+$, $[RNH_3]^+$ and $[R_2NH_2]^+$ there is also a possibility of forming hydrogen bonds to appropriate donors. In fact, the crown ethers have been found to form complexes with these ions in which spatially directed hydrogen bonds determine the specificity. With 18-crown-6, three of the oxygen atoms can participate in hydrogen bonding to $[NH_4]^+$ or $[RNH_3]^+$ cations (Fig. 7.1). The complexes are quite stable and may be isolated in the solid state or as solution species. The determined stability constants for the complexes of ammonium ions with 18-crown-6 in methanol have lg K values in the region of 4 ($[NH_4]I$, lg K = 4.27; $[MeNH_3]I$, lg K = 4.25; $[EtNH_3]I$, lg K = 3.99; $[PhNH_3]Br$, lg K = 3.80). It is even possible to use additional hydrogen bond acceptors in side-chains to provide a fourth binding site for the $[NH_4]^+$ cation, **7.1**. A simple extension to bis(crown)ethers such as **7.2** allows the binding of biologically important polyammonium ions of the type $[H_3N(CH_2)_nNH_3]^{2+}$.

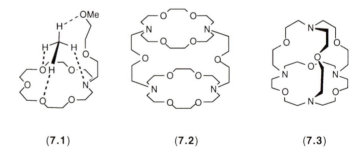

(7.1)   (7.2)   (7.3)

Once three dimensional ligands are envisaged, it is possible to design an encapsulating ligand for the $[NH_4]^+$ cation such that all of the hydrogen bonds are endocyclic. This is achieved in the ligand **7.3** which is known as a *spherand* and which binds $[NH_4]^+$ within the cavity (Fig. 7.2). It is not essential to use nitrogen atoms as the hydrogen bond acceptors and even the cryptand [2.2.2] binds ammonium ions. Interestingly, the binding of $[NH_4]^+$

to [2.2.2] is less precisely defined than with **7.3** and the cation can rotate within the cavity of the cryptand.

Crown ethers and cryptands are being investigated for the selective binding and transport of ammonium compounds in the same way that have been used for the Group 1 and Group 2 metal ions. One unusual application is seen in the enantioselective binding of one of the two enantiomers of chiral ammonium salts $[R^*NH_3]^+$ (where R* is a chiral alkyl residue) by chiral crown ether derivatives such as **7.4**.

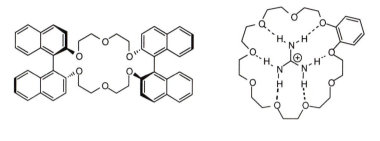

(**7.4**)   (**7.6**)

Other cations may also be bound by macrocyclic ligands and this unusual area of coordination chemistry is likely to increase in the future. Typical examples of hydrogen-bonded interactions include those of $[H_3O]^+$ with **7.5** in which three of the oxygen donors participate in hydrogen bonding and the guanidinium complex **7.6** with benzo-27-crown-9. In this latter adduct, six of the nine oxygen atoms are involved in hydrogen bonding.

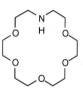

(**7.5**)

7.2 Binding of anions

It is also possible to expand our definition of coordination chemistry to include the design of ligands for anions. Once again, the principal interactions that we will use are based upon hydrogen bonding. However, in contrast to the binding of organic cations, where the ligand acted as the hydrogen bond acceptor and the guest as the hydrogen bond donor, the situation is usually reversed in anion binding. Furthermore, in order to maximize electrostatic interactions it will be beneficial to have cationic hosts and cyclic polyammonium salts (that is to say, protonated polyaza macrocycles and cryptands) fulfil these requirements.

For simple monatomic anions such as chloride the ideal hosts would be spherical and a good example is found with ligand **7.3**. This ligand acted as a hydrogen bond acceptor and formed four hydrogen bonds with $[NH_4]^+$. However, in its protonated form $[H_47.3]^{4+}$ it functions as a hydrogen bond donor and binds chloride ion. However, much simpler hosts may be used and representative examples include **7.7** and **7.8**.

It is possible to design ligands with selectivity for particular halide ions based upon their coordination preferences. The cryptand **7.9** may be hexaprotonated at the secondary amines and the hexacation forms 1:1 complexes with fluoride, chloride and bromide anions. The coordination geometry about the larger chloride and bromide ions is octahedral with six hydrogen bond contacts between the halide ion and the ammonium N-H. However, the smaller fluoride ion only hydrogen bonds to a total of four N-H groups, three from one end and one from the other end of the cryptand (Fig. 7.3).

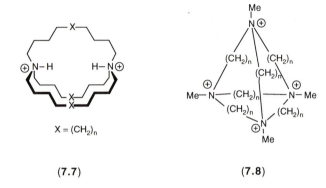

X = (CH$_2$)$_n$

(7.7)   (7.8)

The cryptand **7.9** has an elongated shape that persists after hexaprotonation and this suggested that it might be possible to bind linear anions and this was demonstrated by the formation of a 1:1 adduct in which an N$_3^-$ azide anion is bound within the cavity.

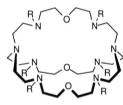

(7.9)

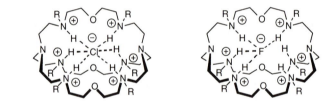

Fig. 7.3. The binding of chloride and fluoride ions within the cavity of the cryptand **7.9**.

Particular interest has focused upon the binding and molecular recognition of carboxylates and phosphates and numerous ligands have been designed to optimize interactions with these anions. Cations derived from simple polyaza macrocycles such as **7.10** and **7.11** have proved to be effective in binding a range of such anions.

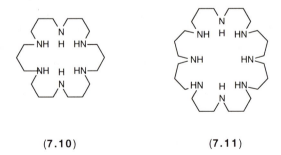

(7.10)   (7.11)

It has even been possible to use the binding of phosphates to macrocycles of this type to develop catalysts for the hydrolysis of adenosine triphosphate (ATP), a process of obvious interest in furthering our understanding of biological phosphoryl transfer processes. The hexaprotonated form of the azacrown **7.12** has proved to be an effective catalyst for the hydrolysis of ATP and other polyphosphates. It is thought that the macrocycle plays a number of roles; firstly, hydrogen bonds from the ammonium groups to the phosphate oxygen atoms serve both to bind the polyphosphate, to activate the ligand towards nucleophilic attack and to bind it in a conformation such that attack by one of the ligand nitrogen atoms becomes possible. A schematic view of the bound polyphosphate is give in **7.13** (charges have been omitted for clarity).

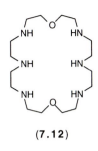

(7.12)

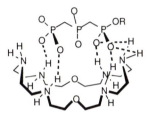

(7.13)

7.3 Biocatalysis by macrocyclic complexes

The high thermodynamic and kinetic stability of macrocyclic complexes of transition metals has resulted in considerable interest in their use as catalysts. In particular, ligands such as cyclam offer the opportunity of tuning the redox properties of metal centres through ligand substitution and also leaving two axial sites available for the binding of substrates.

Numerous catalytic processes based upon transition metal complexes of porphyrin and phthalocyanine ligands have been described. These ligands are particularly important in oxidation reactions involving molecular dioxygen or other simple oxidizing agents and very often involve iron, manganese or cobalt complexes. Much of this work has been directed towards an understanding of the functioning of porphyrins in biological systems.

A few examples of synthetic catalytic systems will serve to emphasize the utility of porphyrin ligands. A variety of iron porphyrin complexes, including very simple systems such as [Fe(tpp)Cl] (tpp = *meso*-tetraphenylporphyrin) are effective in the oxidation of alkanes, alkenes or arenes using PhIO as the primary oxidant. Alkanes are converted primarily to alcohols whilst alkenes give mixtures of epoxides and allylic alcohols. The key step in the reaction is thought to be the conversion of the iron(III) starting material to a formal iron(V) (actually an iron(IV) porphyrin cation) oxo species (Eqn. 7.1).

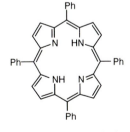

meso-tetraphenylporphyrin

$$[Fe^{III}(L)Cl] + PhIO = [Fe^{IV}(L^{+\cdot})] + PhI \qquad (7.1)$$

Analogous reactions with manganese(III) and molybdenum(V) porphyrins are also known.

However, a number of simpler ligands have also been used in attempts to reproduce biological catalysis and a few examples will be given here. Further examples are to be found in the companion text *Biocoordination Chemistry*. The enzyme carbonic anhydrase catalyses the hydration of carbon dioxide (Eqn. 7.2) and contains zinc(II) at the active site. The coordination number of the zinc is thought to vary between four and five in the catalytic cycle and one of the ligands is a water molecule which is deprotonated with a pK_a close to seven under physiological conditions.

$$CO_2 + H_2O = HCO_3^- + H^+ \tag{7.2}$$

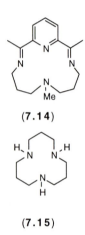

(7.14)

(7.15)

Complexes of the ligand **7.14** are obtained from template condensations of 2,6-diacetylpyridine with $H_2N(CH_2)_3NMe(CH_2)_3NH_2$ in the presence of metal ions. The zinc complex $[Zn(\mathbf{7.14})(H_2O)]^{2+}$ contains a five coordinate zinc centre with one axial water molecule. The pK_a of the coordinated water molecule closely resembles that of the metalloenzyme and the complex shows very modest catalytic activity for the reaction in Eqn. 7.2. Zinc complexes of the macrocycle **7.15** have also been used as models for the active site of carbonic anhydrase and, although they are not particularly effective for the hydration reaction they reproduce a number of the dynamic features associated with a change in coordination number from four to five and also interact with thiocyanate SCN^-. This latter ion is frequently used as a model for carbon dioxide as it possesses the same linear geometry.

The structure of Vitamin B_{12} was introduced in Chapter 2. At the time of its structural elucidation it was thought remarkable that a cobalt(III) complex could form stable Co-C bonds in aqueous solution under physiological conditions. It was thought that the macrocyclic ligand was important and numerous cobalt(III) macrocyclic complexes were prepared in attempts to reproduce this behaviour. The studies proved to be rather too successful – the vast majority of macrocyclic ligands are able to give complexes containing stable Co^{III}-C bonds, but in retrospect it turns out that the majority of non-cyclic ligands are also capable of such stabilization. Nevertheless, there is a large variety of macrocyclic derivatives containing organometallic cobalt(III) centres which reproduce some of the features of Vitamin B_{12}.

Hemerythrin and hemocyanin are both dioxygen carrier proteins found in various types of invertebrates. The dioxygen-binding sites contain two iron or two copper centres respectively. Much effort has been expended in the preparation of model dinuclear complexes which reproduce the spectroscopic (if not the chemical) properties of these metalloproteins. Some of the most successful models have used the very simple ligand triazacyclononane and Fig. 7.4 shows a dinuclear iron complex of this ligand which reproduces many of the structural features of haemerythrin. The key features are the presence of two bridging carboxylate ligands together with a bridging oxo ligand. More recent work with triazacyclononane has provided models for a range of other iron and copper metalloproteins. Some derivatives of triazacyclononane have found commercial application as ligands in manganese-based oxidation catalysts.

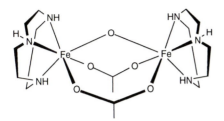

Fig. 7.4. A complex of triazacyclononane that reproduces features of the active site of hemerythrin.

We conclude this section by noting a number of additional biological processes in which metal complexes of macrocyclic ligands are involved. Because so many biological processes involve iron complexes, very sophisticated mechanisms have evolved for the control of iron metabolism. Some aerobic bacteria have developed extremely effective ligands known as *siderophores* for the binding of iron. These iron(III) complexes of these ligands are very stable (lg $K > 30$). The ferrichromes are one class of siderophore and consist of 18-membered ring oligopeptide macrocycles bearing appended hydroxamate groups. Three deprotonated hydroxamate groups bind the iron as iron(III) and are so effective that the bacteria can leach iron from stainless steel. A different macrocyclic ligand called enterobactin is used by enteric bacteria. Once again amino acids provide the macrocyclic structure, but in enterobactin the iron is bound by three catechol substituents. This compound has the highest known binding constant for iron(III) with a lg K value in the region of 52! Because the macrocyclic core is derived from (chiral) amino acids, the absolute stereochemistry at the metal centre (Δ or Λ) is predetermined. Naturally occurring enterobactin gives a Δ iron complex.

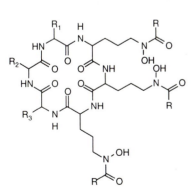

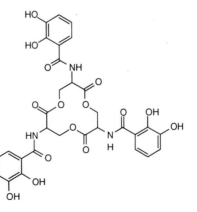

ferrichrome

enterobactin

7.4 Concluding remarks.

This book has given no more than an overview of macrocyclic coordination chemistry. Some indication of the range of ligands which have been studied has been given, together with key features of the structural and reaction chemistry of the complexes that are formed. Throughout the book, various applications have been briefly discussed although these have not been collected together into a single chapter.

It is probably true that macrocyclic chemistry is currently in a consolidation phase. The basic synthetic strategies are well-established and the coordination chemistry is well-understood. Much of the current effort in the area is related to optimizing properties of interest and the fabrication of devices incorporating macrocyclic ligands.

Is macrocyclic chemistry any different from other areas of coordination chemistry? Probably not - but it has been responsible for a change in the way in which inorganic chemists think about the subject.

And as I said at the beginning, macrocyclic chemistry is, above all, fun.

Further reading

The textbooks and monographs listed below contain good accounts of macrocyclic coordination chemistry. Most contemporary textbooks of inorganic or bioinorganic chemistry also contain chapters or significant sections on macrocyclic ligands.

Bernal, I. ed. (1987). *Stereochemical and Stereophysical Behaviour of Macrocycles*, Elsevier, Amsterdam.

Black, D.St.C. (1987). Chapters 7.4 and 61.1 in *Comprehensive Co-ordination Chemistry*, eds. G. Wilkinson, R.D. Gillard and J.A. McCleverty, Pergamon, Oxford.

Constable, E.C. (1995). *Metals and Ligand Reactivity*, VCH, Weinheim.

Dietrich, B., Viout, P., and Lehn, J.-M. (1993). *Macrocyclic Chemistry*, VCH, Weinheim.

Fenton, D. (1995). *Biocoordination Chemistry*, Oxford University Press, Oxford.

Gutsche, C.D. (1989). *Calixarenes,* RSC, Cambridge.

Gokel, G.W. (1991). *Crown Ethers and Cryptands*, RSC, Cambridge.

de Sousa Healy, M. and Rest, A.J. (1978). *Adv. Inorg. Chem. Radiochem.*, **21**, 1.

Henrick, K., Tasker, P.A., and Lindoy, L.F. (1985). *Progr. Inorg. Chem.*, **33**, 1.

Lehn, J.-M. (1995). *Supramolecular Chemistry*, VCH, Weinheim.

Lindoy, L.F. (1989). *The Chemistry of Macrocyclic Ligand Complexes*, Cambridge University Press, Cambridge.

Izatt, R.M., Christensen, J.J. eds. (1978). *Synthetic Multidentate Macrocyclic Compounds*, Academic Press, New York.

Melson, G.A. ed. (1979). *Co-ordination Chemistry of Macrocyclic Compounds*, Plenum, New York.

Newkome, G.R. ed. (1997). *Comprehensive Heterocyclic Chemistry II*, Vol. 9. Elsevier, Oxford.

Parker, D. (1996). Macrocycle Synthesis, A Practical Approach. OUP, Oxford, 1996.

Vögtle, F. and Weber, E. (1985). *Host Guest Complex Chemistry*, Springer, Berlin.

Index